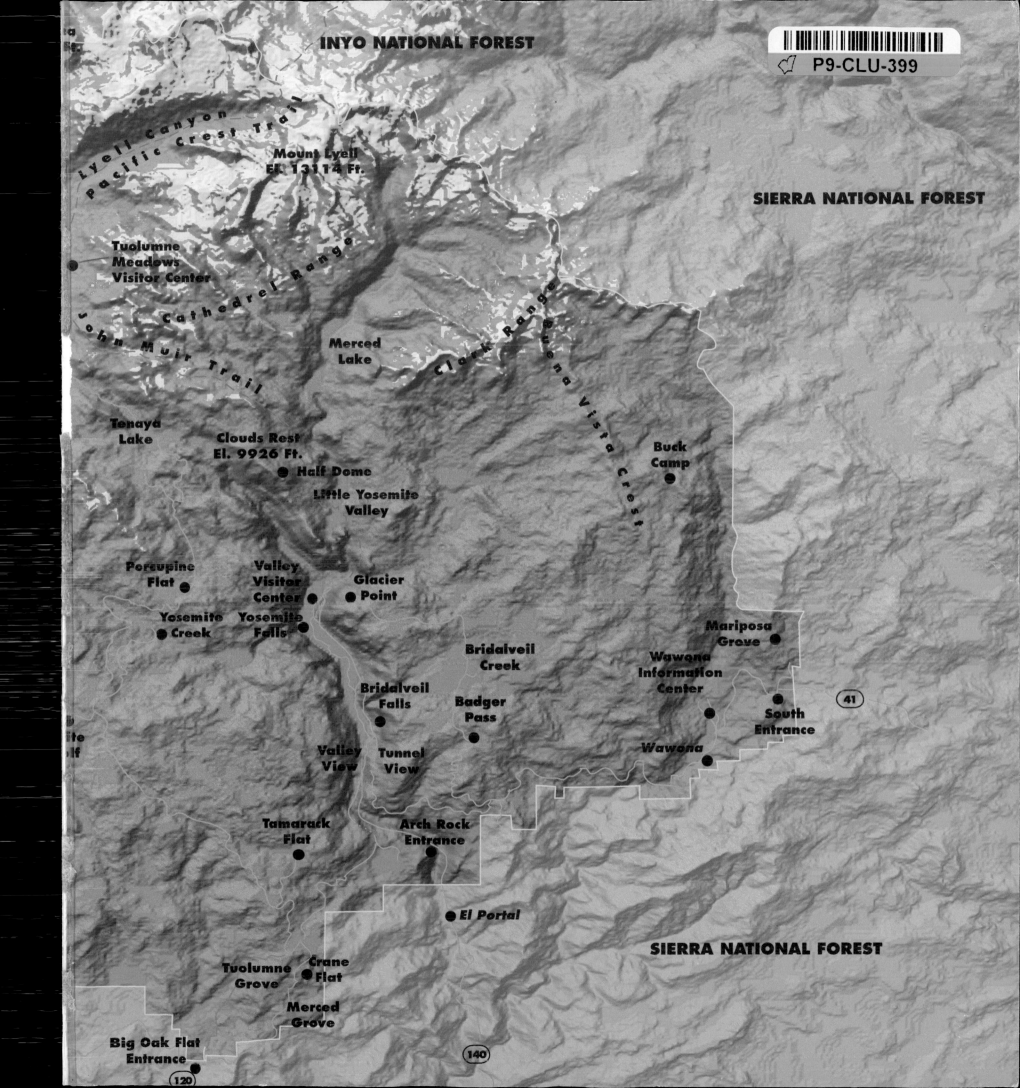

INYO NATIONAL FOREST

P9-CLU-399

SIERRA NATIONAL FOREST

Lyell Canyon

Pacific Crest Trail

Mount Lyell
El. 13114 Ft.

Tuolumne
Meadows
Visitor Center

Cathedral Range

John Muir Trail

Merced
Lake

Clark Range

Buena Vista Crest

Tenaya
Lake

Clouds Rest
El. 9926 Ft.

Buck
Camp

Half Dome

Little Yosemite
Valley

Porcupine
Flat

Valley
Visitor
Center

Glacier
Point

Yosemite
Creek

Yosemite
Falls

Mariposa
Grove

Bridalveil
Creek

Wawona
Information
Center

41

Bridalveil
Falls

Badger
Pass

South
Entrance

Valley
View

Tunnel
View

Wawona

Tamarack
Flat

Arch Rock
Entrance

El Portal

SIERRA NATIONAL FOREST

Tuolumne
Grove

Crane
Flat

Merced
Grove

Big Oak Flat
Entrance

140

120

"Waning sun, shuttered behind a pure white cloud, traces a rim with eye-blinding incandescence. . . . Tall, dark green, narrow triangles of trees rise against a backdrop of white granite beneath a deep blue Sierran sky: unmistakably Yosemite."

ANN ZWINGER

YOSE

VALLEY OF

ANN

PHOTOGRAPHY BY

HarperCollins*San Francisco* *A Division of HarperCollinsPublishers*

MITE

THUNDER

ZWINGER

KATHLEEN NORRIS COOK

A Tehabi Book

The Genesis Series was conceived by Tehabi Books and published by Harper Collins San Francisco. The series celebrates the epic geologic processes that created and continue to shape America's magnificent national parks and their distinctive ecosystems. Each book is written by one of the nation's most evocative nature writers and features images from some of the best nature photographers in the world.

Yosemite: Valley of Thunder was produced by Tehabi Books. Nancy Cash–*Managing Editor*; Andy Lewis–*Art Director*; Sam Lewis–*Art Director*; Tom Lewis–*Editorial and Design Director*; Sharon Lewis–*Controller*; Chris Capen–*President*. Additional support for *Yosemite: Valley of Thunder* was also provided by Susan Wels–*Series Editor*; Jeff Cambell–*Copy Editor*, and Anne Hayes–*Copy Proofer*.

Written by Ann Zwinger, *Yosemite: Valley of Thunder* features the photography of Kathleen Norris Cook. Supplemental wildlife photography was provided by Michael Frye (pages 85, 90-91, 92, 93a, 93c, 94a, 94c, 94d, 95a, 95b, 95c, 96, and 100-101). Technical, 3-D illustrations were produced by Sam Lewis. Source materials for the illustrations were provided as digital elevation models from the United States Geological Survey. Additional illustrations were produced by Andy Lewis and Tom Lewis.

For more information on Yosemite National Park, Harper Collins San Francisco and Tehabi Books encourage readers to contact the Yosemite Association, P.O. Box 545, Yosemite National Park, CA, 95389; (209) 379-2646.

Harper Collins San Francisco and Tehabi Books, in association with The Basic Foundation, a not-for-profit organization whose primary mission is reforestation, will facilitate the planting of two trees for every one tree used in the manufacture of this book.

Library of Congress Cataloging-in-Publication Data
Zwinger, Ann
    Yosemite: valley of thunder / Ann Zwinger : photography by
Kathleen Norris Cook. — 1st. ed.
        p. cm.—(The Genesis Series)
    "A Tehabi book"
    Includes index.
    ISBN 0-06-258570-3 (cloth).—ISBN 0-06-258561-4 (pbk.).
    1. Yosemite National Park (Calif.) 2. Yosemite National Park
(Calif.)—Pictorial works. 3. Natural History—California—Yosemite
National Park. 4. Natural History—California—Yosemite National
Park.—Pictorial works. 5. Geology—California—Yosemite National
Park. 6. Geology—California—Yosemite National Park—Pictorial works.
I. Cook, Kathleen Norris. II. Title. III. Series: Genesis Series.
F868.Y6Z96
979. 4'47—dc20   95-47251
                                                          CIP

96 97 98 99 TBI 10 9 8 7 6 5 4 3 2 1

This edition is printed on acid-free paper that meets the American National Standards Institute Z39.48 Standard.

PAGE 1:

*M*ist rises *from El Capitan.*

PAGES 2-3:

Yosemite Falls seen from the South Rim at Taft Point.

PAGES 4-5:

The Leaning Tower rises above the South Rim of Yosemite Valley.

PAGES 6-7:

Looking across Yosemite Valley, the mists of Bridalveil Fall blend with those departing after a spring storm.

PAGES 8-9:

Yosemite Falls plunges 2,425 spectacular feet to the valley floor.

THE GENESIS SERIES
YOSEMITE
VALLEY OF THUNDER

Granite Faces, Dusty Trails: The Legacy of Discovery   *13*

Gatefold: Images of Yosemite Valley   26

Heart of Stone: Domes, Glaciers, and River Valleys   *35*

Gatefold: Topography of Yosemite   44

Triumph of Trees: Plant Life of Wood, Slope, and Meadow   *59*

Gatefold: The Forests of Yosemite   74

Life on Wing and Water: Wildlife of the Valley   *79*

Gatefold: Animal Communities—The Diversity of Life   92

Walks on the Wild Side: Along Streams, Springs, and Falls   *103*

Trumpets of Light: Backcountry Journal   *117*

Acknowledgments   *131*
Index   *132*

I crest the path at Glacier Point and gasp: before me is the astounding and magnificent stage set of Yosemite Valley, grander, more monumental, than any painting or any photograph I have seen or any verbal description I have read. Over three thousand feet below me, the valley spreads out in full panoply, swathed in the honeyed light of late afternoon.

Deep green trees, repeated narrow verticals, blanket the valley. The sleek, elegant, pearl gray granite of Half Dome soars out of the deep, light-absorbent greens. Its clean-sliced wall dominates the valley, perfectly displayed from this angle. This panorama contains the quintessential juxtaposition of Yosemite, the cool curves of granite domes against dark, serrated horizons of trees in their "unfading luxuriance," as Horace Greeley put it, writing about Yosemite. Their somberness frames and displays the domes and cliffs like gems on a jeweler's velvet. The massive, near-white granites, billowing in a succession of domes at every turning, eclipse everything else in the landscape, even the two waterfalls that animate the walls below.

I open my beautiful, brand-new topographical map, its folds still reluctant. On it, Yosemite is set like a cameo in the necklace of the Sierra Nevada. The park itself is shaped like a huge bear paw, appropriate enough since Yosemite is

Cathedral rocks rise from the valley floor.

GRANITE FACES, DUSTY TRAILS
THE LEGACY OF DISCOVERY

thought to mean "grizzly." I trace the boundaries with my finger, scan the landmarks in front of me, identify domes and snow-patched mountains and waterfalls, locate where I am, where I've been, and where I'm going.

Late sunlight smoothes the rock faces with amethyst light, rounding edges, creating massive sculptures. For a moment Half Dome glows with a remarkable incandescence of color that persists, attenuated, as if for this extraordinary landscape the sun stands still and blesses the scene with an extra benediction of light. Then daylight fades, mauve segues to silver, and the valley trembles with darkness.

Reading them later, my notes of this first trip remind me of weariness and traveling all day, too many people crowding the paths, a smelly outhouse, and hundreds of yellow jackets to which I am violently allergic buzzing everywhere. But I remember only this: the overwhelming first view of Yosemite Valley that gave me serenity of mind and spirit and the pure, astonishing beauty of it all.

* * *

I trace on the map the two trails that connect Glacier Point to Yosemite Valley below. One is the old "Four-Mile Trail," which is now a five-mile trail because of alterations to take out some of the steepness. John Conway, who built several trails in Yosemite, bossed the Four-Mile Trail to the valley during 1871–72 for James McCauley, who built "Mountain House" here on Glacier Point in 1878. It's likely that McCauley instituted Yosemite's "firefalls" to entertain tourists one Fourth of July in the early 1870s. The fifteen-hundred-foot freefall of glowing coals and burning boughs shoved off Glacier Point was stopped in 1968 after it was deemed incompatible with park philosophy. Friends of mine who went to Yosemite as children still remember these firefalls as awesomely beautiful.

I follow the second trail, the much longer Pohono Trail, a track first trod by the Ahwahneechee Indians and from which the first Euro-American visitors initially viewed the valley, because I want to see what they saw.

The very first white man to see the valley was probably William Penn Abrams, who became lost when he and a companion were chasing a grizzly bear. His diary entry for October 18, 1849, recounted that they followed an Indian trail

> . . .that led past a valley enclosed by stupendous cliffs rising perhaps 3,000 feet from
> their base and which gave us cause for wonder. Not far off a waterfall dropped from a
> cliff below three jagged peaks into the valley, while farther beyond a rounded moun-
> tain stood, the valley side of which looked as though it had been sliced with a knife as
> one would slice a loaf of bread.

Two years later, James Savage, whose trading post on the Merced River had been ransacked by the same Ahwahneechee, assumed leadership of the all-volunteer Mariposa Battalion, which had been organized to handle the Indian "problem." The Ahwahneechee had

Seen from the Wawona Tunnel exit, fog swirls in Yosemite Valley while snow garnishes Cloud's Rest.

inhabited the valley for some thirty-five hundred years, had harvested its acorns to make gruel and cakes, woven their baskets from its rushes and willows, and burned its meadows to encourage the growth of black oaks and berry-bearing shrubs. Distressed by the influx of white settlers— whom they felt, with justice, were usurping the lands from which they gathered their food—the Ahwahneechee had taken to harassing local settlements; hence the expedition.

Late on the afternoon of March 27, 1851, Savage led a detachment of fifty-eight men down into the valley. Among them was Lieutenant Lafayette Houghton Bunnell, then twenty-seven. Until the discovery of Abrams's diary in 1947, Bunnell was assumed to be the first Euro-American to record his impressions of the panorama. On the descent, the men passed a promontory at 6,640 feet, now called Old Inspiration Point, where the valley opened before them. Most of the soldiers were in a hurry to get on down to the bottom, but Bunnell, a literate and observant man who remained devoted to Yosemite, halted his horse at the edge of the cliff to absorb the view.

Once in the valley, the battalion found few Indians. The majority had cleared the valley and gone eastward. The men camped in the meadow near an exquisite fall, later named Bridalveil by someone with overly sentimental nineteenth-century sensibilities. Around the campfire, Bunnell proposed they name the valley "Yo-sem-i-ty" to commemorate the Indians they were dispossessing— thinking, mistakenly, that was what the Indians called themselves.

The Mariposa Battalion departed at the end of six months, their tour of duty ended. The Ahwahneechees returned as soon as the troops were gone. The next year, when more miners were killed in the vicinity, another U.S. Army group persuaded the Indians to leave permanently. The Ahwahneechee chief, Tenaya, after whom several landmarks in the valley are named, was killed in 1853 by another Indian tribe. After his death the Ahwahneechee scattered and never reorganized.

Bunnell, who wrote a popular book entitled *Discovery of the Yosemite* twenty years later, admitted that the view of the valley never struck him again with the same force and poignancy that it had on his first encounter, and he bemoaned the fact that later visitors, already familiar with Yosemite's sights through written accounts and paintings, never had the privilege of that first astounding, pristine view. I at first agreed with Bunnell. But I have since come to feel that no matter how many portrayals one peruses before coming, from Thomas Moran to John Muir to Ansel Adams, the valley's physical presence is so overwhelming, so commanding and so powerful, that each of us has our own memorable first view.

* * *

One chill May day, I start at the base of the old Pohono Trail to hike up to the two "Inspiration Points" that commemorate the two first views of the valley. This morning, most of the path is a rivulet, and waterfalls punctuate the north slope of the valley. Thin,

In a high, lush Sierran meadow, Indian paint-brush surrounds sweet-cicoly, thistles, fleabane daisies, and tall buckwheat.

B*right yellow sulphur flowers, a member of the Buckwheat family, rise above their furred, spoon-shaped, gray-green leaves.*

W*ildcat Falls cascades down the cliffside in the springtime.*

Half Dome seems to levitate above the vapors of a spring snow that dusted the high ridges with snow.

FOLLOWING PAGES:

The Leaning Tower
catches the last glow of
afternoon while shadows
fold into Bridalveil
Fall—called "Po-ho-no"
or "Fall of the Puffing
Winds" by prehistoric
Indians.

*B*oulders, placed
as if with an eye to
balance and harmony,
accent a reed-sprigged
pool reflecting sunset.

feathery threads, sometimes straight, sometimes braided, chalky white, peel off the hillsides and slather down cliff faces.

Early stagecoach passengers first saw the valley from the lower Inspiration Point. What was once a handsome vista is now interrupted and partially obscured by tall snags and a big Douglas fir whose branches crosshatch the view. I wonder how on earth a stagecoach ever came bucketing down this road without launching half a dozen terrified passengers off into space.

The trail continues upward through the mixed conifer forest that is so much a trademark of Yosemite, an amalgam of lodgepole and big Jeffrey pines, white cedar, sugar pines, and Douglas firs. In this spring of heavy snows, several trees have toppled across the path. Lichens colonize the bole and branches of a newly fallen incense cedar: pale gray lichen, flat and crusty; leafy lichen with curly edges; fluorescent green staghorn lichen with hollow branching stalks; and dark brown horsehair lichens that ruffle its branches.

Fifteen hundred feet farther up trail, bespattered with mariposa lilies and thimbleberry in bloom, is Old Inspiration Point. From this lookout, artist Thomas Ayres made the first sketch of the valley in 1855, having been brought here by James Hutchings, editor and publisher of *The California Illustrated Magazine*, as well as promoter and hotelier par excellence of Yosemite. It was from here also that Bunnell recalled his moment of apotheosis:

> *The grandeur of the scene was but softened by the haze that hung over the valley—light as gossamer—and by the clouds which partially dimmed the higher cliffs and mountains. This obscurity of vision but increased the awe with which I beheld it, and as I looked, a peculiar exalted sensation seemed to fill my whole being, and I found my eyes in tears with emotion.*

* * *

I doubt that Bunnell, astride his horse on that blowy March afternoon when he first glimpsed the pristine valley, ever guessed how quickly commercial development would invade the area. Hotel and trail building proceeded apace in the years after Yosemite's discovery, but the remoteness of the region and the difficulty of access kept visitor numbers low. Between 1855 and 1864, only 653 visitors registered at Yosemite hostelries.

Although overnight visitors were few, there were growing signs of misuse by the promoters, who farmed and ranched on the valley floor to support the hotels they were building. Unfortunately, a huge part of the Sierra Nevada, including what would become Yosemite, were unsurveyed federal lands, and the federal government had no means to "protect" such territory until 1864, when President Abraham Lincoln signed an act that ceded to California the Mariposa Grove of Big Trees near the Wawona entrance plus Yosemite Valley, a total of 48.6 square miles. In the first instance of congressional recognition and preservation of land for its scenic values alone, Congress set aside these two tracts "for public use, resort and recreation . . . inalienable for all time."

IMAGES OF YOSEMITE VALLEY

Artists were delighted with Yosemite Valley from the beginning. From the first rapturous portrayals of Thomas Ayres in 1855 to the golden-lit atmospheric canvases of Albert Bierstadt in the 1860s, the artist's vision of Yosemite provided the only information about the area to a curious public. The full-blown romanticism of Bierstadt influenced the way easterners perceived the West and would color tourists' reactions to these stunning vistas—which were, in the end, even more magnificent and even more spectacular than the artist's renderings.

Others portrayed Yosemite in words. John Muir came to Yosemite in 1868 and

found an enchantment "quivering with sunshine and lark song." Although he would remain here only six years, he trod and climbed and probed every corner of the park, often carrying only a bag of tea and a loaf of hard bread. Muir translated his enchantment and astonishment at the colors and sounds and forms of Yosemite into a flowing eloquence: "No temple made with hands can compare with Yosemite, the grandest of all the special temples of Nature." Muir's devotion to Yosemite had far-reaching consequences. In establishing the Sierra Club in 1892, Muir set forth the philosophy of conservation that ballasts the thinking of all those who believe in the validity of wilderness.

Many who have traveled through Yosemite—including volunteer soldier Lafayette Houghton Bunnell, politician Teddy Roosevelt, newspaperman Horace Greeley, scientist Joseph LeConte, and photographer Ansel Adams—have left telling words that witness their own captivation with the wild, majestic beauty of the place. But none of those words has been as compelling and consequential as those written by John Muir.

John Muir

When he received news of Lincoln's signature, the governor of California appointed the Yosemite Commission to oversee matters for Yosemite Valley and the Mariposa Grove. It was a blue-ribbon group of respected men, chaired by the well-known landscape architect Frederick Law Olmsted. Olmsted insisted on keeping the park a "pure sanctuary," sensitive to what large numbers of visitors could do in the way of damage: "An injury to the scenery so slight that it may be unheeded by any visitor now, will be one of deplorable magnitude when its effect upon each visitor's enjoyment is multiplied by these millions."

Olmsted's concern was prescient. In 1864, 147 tourists visited Yosemite. In 1994, 4.4 million tourists arrived, on occasion filling every single parking lot in the park. Olmsted, however, resigned from the commission in October, 1866, to go to New York to oversee his design of Central Park, and his authoritative report, waylaid by prodevelopment interests, never reached the legislature. After Olmsted's departure, men staffed the commission who saw no dichotomy between protecting the park and encouraging commercial interests. They believed the prime purpose of the park was to provide food, lodging, and "entertainment" for visitors, as well as steady income for concessionaires. The conflict still endures today.

* * *

Among the 627 people who traveled to Yosemite in 1868 was a botanist and self-declared naturalist, John Muir. He walked off the ship in San Francisco and, having read Hutchings's descriptions of the Sierra Nevada, headed east. He found hillsides vibrating with the color of spring wildflowers and the air "quivering with sunshine and lark song." His first view of the snow-bright phalanx of the Sierra Nevada struck him like a homecoming. Although he remained in Yosemite a relatively few years, the landscape imprinted on him like a mother goose on its gosling. The longer he stayed, the more the mountains provided sustenance and challenge for his lively mind and receptive psyche. The power of Yosemite, combined with Muir's particular character, engendered ideas and words that have influenced the nation's thinking about wilderness ever since, and Yosemite is forever endowed with his spirit.

Almost as soon as he arrived in the park, Muir began lobbying for the return of Yosemite Valley and the Mariposa Grove to federal jurisdiction and protection under federal law, since California, via the Yosemite Commission, had proved mainly interested in development.

However, returning Yosemite to federal jurisdiction in 1905 (with the exception of Yosemite Valley and the Mariposa Grove) did not solve the problem because Yosemite, like Yellowstone, was without any provision for administration or protection. The National Park Act, which would provide protection, was not passed until 1916. In 1891 troopers of the Fifth U. S. Cavalry arrived to provide supervision. They set up camp in Wawona and patrolled the park, except for the two plots still under state jurisdiction.

In 1892, Muir gathered a small group of men to pursue sequestering Yosemite in toto. The group evolved into the Sierra Club. As president of the new association, Muir

These stand of black oak is a remnant of the capacious groves that once grew in Yosemite Valley and provided an essential food supply for the valley's inhabitants. Prehistoric Indians set fires from time immemorial to destroy competing trees and encourage the growth of black oaks. Today, fire control policies have nearly decimated the once-lovely groves.

bent Theodore Roosevelt's ear about the amount of development in the valley, ticking off how saloons, post office, general store, hotels, stables, barns, art galleries, and private dwellings were impinging on its beauty (today there are thirteen hundred buildings, plus a courthouse and jail, and seventeen acres of parking lots). In 1905, after Muir's politicking, the State of California re-ceded Yosemite Valley and the Mariposa Grove to the control of the federal government. In so doing, the overall acreage was reduced, and Yosemite became essentially the park it is today.

The cavalry left in 1914, replaced by civilian rangers. The troops had already pioneered a good system of trails and mapped the park in substantial detail. They had also virtually eliminated grazing by the simple solution of escorting an apprehended herder to the farthest limits of the park. By the time he rejoined his flock, days later, the sheep had usually dispersed or been gobbled up by bears or mountain lions. Hikers today will still come across large "T" blazes on trees, the old trail mark used by the Fifth Cavalry to guide their travel through the park. Standing beneath one, I imagine the cavalry a century ago, bringing order out of chaos, leaving these markers for posterity.

The departure of the cavalry was followed by congressional establishment of a national park system in August 1916, two years after Muir's death. For the first time, a federal department was given the sole responsibility of caring for designated wilderness lands.

* * *

In 1899, the year before the first automobile drove into Yosemite, two schoolteachers from Indiana arrived at the park, the abrasive David C. Curry and his wife, Jennie. The couple began their tourist business by pitching seven tents at the foot of Glacier Point, where Curry Village is still located. Curry consistently alienated every park superintendent by going directly to the press to get what he wanted, continually expanded his position by fair means and foul, and reaffirmed the concessionaire philosophy that luxuries are where the profits are, and money is to be made by making luxuries into necessities. Curry was less successful when he proposed, among other things, damming Yosemite Creek above the falls. The dam would hold back water that could be poured over the cliff when the falls normally dried up in the summer. Otherwise, his business thrived. By the time Curry died in 1917, Curry Camps welcomed twenty-five to thirty thousand guests a year, a fivefold increase in fifteen years. They continued and vastly expanded the development James Hutchings had begun in the 1860s, setting the tone for development in Yosemite that still exists today.

Then as now, the heaviest visitor pressure is in Yosemite Valley. Yosemite has the largest overnight visitor list and police and medical requirements of any of the national parks. Its staff includes more than 750 full-time and seasonal workers; concessionaires add 1,800 workers in the summer.

"Commercialism" is the only word for what exists in Yosemite Valley today, and the Park Service does as well as possible in handling huge crowds of people. The low point was when a mob of unruly campers rioted on July 4, 1970. Theft and vandalism continue. In the last few years, on big holiday weekends, when head counts have indicated that Yosemite has no more parking spaces, the park has closed its entrances, to the very vocal consternation of concessionaires inside and outside the park. Despite the pressure of crowds and the demands of law enforcement, the park manages to provide superb interpretation and services.

      Yosemite crystallizes the two opposing points of view concerning the role of national parks. Olmsted's prophecy has painfully come true. Should we preserve the parks as natural ecosystems and open-air classrooms or increase opportunities for development? Yosemite most poignantly poses that dilemma.

When I stand on Sentinel Dome's bare pate, some four thousand feet above the valley's floor, I have an incomparable 360-degree panorama as well as exposure to a brisk, jagged wind that also comes from every direction. I hunker down behind a boulder and try to visualize what Yosemite looked like once upon a time and long, long ago. In my mind's eye I watch ancient sediments pour into the ocean and tectonic plates bump and grind. I imagine pale granites forming beneath the surface like immense mushrooms and cold, silver cords of ice strangling the land.

The area that would be called Yosemite lay at the western edge of the expanding continent of the North American plate. The oldest rocks in Yosemite began as sediment deposited in a vast sea hundreds of millions of years ago. The old ocean's shore lay well east of the present Pacific Coast, possibly near what is now Nevada. Rain-loosened and gravity-pulled sediments of mud, silt, and sand sloughed off highlands to the east and poured into the ocean, where they formed thick deposits.

Under their weight, these sediments compacted into shale and sandstone intermixed with limestone made up of small-shelled ocean creatures. Today, the limestone provides some of the few fossils for dating the earliest millennia of Yosemite. Eventually, these sedimentary rocks were pushed so far down into the hot interior of the earth that heat and pressure recrystallized their mineral grains and converted them

J. S. Chase, an Englishman who made a circuit of the Yosemite Rim in the early 20th century, wrote about the Fissures, the "remarkable, — vertical clefts in the west face of a deep side-cañon" which still present dizzying glimpses down their sheer, narrow slots into unfathomable distances below.

Heart of Stone
Domes, Glaciers, and River Valleys

into metamorphic rocks. Today, less than 5 percent of Yosemite's surface rock is metamorphic, and these outcrops vary greatly in their amount of alteration. Some show the original layering of sandstone, limestone, or shale, and some have been completely crystallized by time, heat, and pressure into other rock types such as hornfels and slate.

* * *

More than a million years ago, the Pacific plate began to slide under the North American plate in a process geologists term "subduction." In the life of the earth, new crust is constantly being created by rising magma, and old crust is constantly being destroyed.

Yosemite's granites are plutons, huge masses of magma that cooled and crystallized five or six miles below the earth's surface. A vast batholith— a cluster of separate plutons—underlies the entire Sierra Nevada, making it the most homogenous mountain range in the United States. In Yosemite, erosion has stripped away the metamorphic rocks that once covered the surface of the plutons, revealing the massive granite domes that are so characteristic of the park.

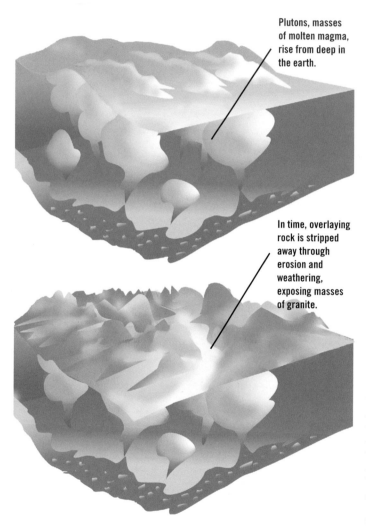

Plutons, masses of molten magma, rise from deep in the earth.

In time, overlaying rock is stripped away through erosion and weathering, exposing masses of granite.

Subduction is the process that adds new material to already existing land masses. Where subduction zones occur along plate margins, massive volumes of magma, generated deep within the earth, rise toward the surface. Magma may erupt in ash- and lava-spewing volcanoes that form atop the overriding plate (the line of volcanoes marks the locus of the subduction zone beneath), or the magma may cool and crystallize far beneath the earth's surface. Yosemite's granites are plutons, huge masses of intrusive igneous rock, that formed five or six miles below the surface. Geologists have identified between fifty and a hundred discrete plutons in the park's area, and they can often be distinguished by their different colors and textures. When separate plutons cluster together, they are termed a batholith. A vast batholith underlies the Sierra Nevada, the longest and most homogenous mountain range in the United States.

When granite cools slowly, it develops crystals large enough to be clearly visible to the human eye (the word "granite" comes from the same root as "grainy"). Granite's salt-and-pepper surface derives from three minerals: quartz, feldspar (two kinds), and the "dark minerals," such as mica or hornblende. The difference in the proportion of these minerals determines the different types of granite. If the proportions of quartz and the feldspars are roughly equal, the rock is pure granite. In Yosemite, most of the granites contain more plagio-clase feldspar (a mineral that is heavy in calcium and sodium) and are technically termed "granodiorites."

Around 80 million years ago, volcanoes stopped spouting and granites stopped forming, and the ancestral Sierra Nevada probably reached heights over 13,000 feet. They must have resembled the Cascade Range in western Oregon and Washington, mountains that are also being built today over an active subduction zone. When volcanic activity stopped, erosion began to shape the mountains, removing the volcanic carapace and nearly all of the old metamorphic rocks, uncovering the underlying pale granite. By 25 million years ago, the Yosemite area had been ground down to a spacious, rolling upland, threaded by modest streams.

Beginning about 25 million years ago, the area of Yosemite was faulted and tilted upward like a huge wedge. West to east, the land slopes upward like a gentle ramp with a 2 to 6 percent grade for fifty miles. At the sharp-breaking Sierra crest, the mountains plummet down more than 6,500 feet to the Mono Basin. Geologists estimate that Mount Dana rose some 6,500 feet to its present-day 13,000 feet in the last 10 million years and may indeed still be increasing in height.

The uplift and tilting of the Sierra steepened stream gradients, increasing their erosive ability, making it possible for them to carry away all of the rock debris weathered out of canyon walls and cut into the valleys to leave deep, V-shaped gorges. In addition to downcutting, streams chiseled away the sides of their channels, removing support from the banks, which then collapsed and also washed downstream.

* * *

Granite cools and crystallizes beneath earth's surface under significant pressure. After erosion removed the massive cap of metamorphic and volcanic rocks and exposed the granite, it reacted to the release of pressure by fracturing along planes of weakness, called joints. Into these joint cracks water seeps, freezes, and expands, a constant mallet-and-wedge process that eventually shatters even the most solidly cemented rock and is visible as elaborate, webbed patterns of dark lines on white rock walls.

However, a great deal of Yosemite's granite is unjointed. It is in these coherent masses that the solid cliffs and the distinctive domes that are so much a part of the Yosemite landscape develop. From the single vantage point of Sentinel Dome's summit, I count thirty granite domes of various sizes, some low and almost tree-covered, others high and commanding, agleam in the sunshine like huge, well-polished newel posts.

These domes can form only in massive, unlayered rock like granite, never in layered sandstones or limestones or in heavily jointed rock. When pressure is removed from a solid monolith that has no joints or other prescribed pathways for release, the rock's outer surface expands more quickly than its core. The outside surfaces exfoliate, "leafing off" patches of rock in curved layers, somewhat resembling the layers of an onion. These "peels" vary in thickness from less than a foot to many feet. As succeeding shells are freed and disintegrate, projecting corners are rounded, and over time the monolith evolves into a more or less smoothly curved dome.

Yosemite's domes, rare elsewhere, were formed and exposed before the onset of glaciation 2 million years ago. Josiah Whitney, State Geologist of California in the mid–nineteenth century, was mistakenly convinced that Half Dome had been "split asunder" and that the missing half had been lost in "the wreck of matter and the crush of worlds." In truth, Half Dome is very nearly as intact as it ever was. During maximum glaciation, Half Dome stood five hundred to seven hundred feet above glacial inundation. It was vertical jointing, not glaciation, that gave Half Dome its sheer face, and exfoliation its rounded backside. While ice did not essentially change many of the domes already formed, it did recontour and polish others up a bit by removing broken shells.

One warm summer morning, I sit on an unnamed dome and look south to Lembert Dome, both perfect examples of domes reshaped by ice, which may have covered these to a depth of one thousand feet. Grinding ice left a gentle slope on the upglacier side, but it plucked out rock on the lee side and left it nearly vertical. Although these glacier-worked landmarks are also called "domes," they are not symmetrical, as Sentinel and North Domes are. They are asymmetrical knobs and hillocks called "roches moutonnées" for their resemblance, when grouped, to the fleecy, crimped barrister wigs worn in French nineteenth-century courts.

* * *

John Muir first proposed that Yosemite Valley had been glaciated, a bold theory since no less a light than Josiah Whitney was convinced that Yosemite Valley had dropped in a series of "cataclysmic events" and that there had been no glaciers in Yosemite Valley. He ungraciously called Muir "a mere sheepherder, an ignoramus." Muir's views received attention after Joseph LeConte, a geology professor at the University of California, visited Yosemite in 1870 and looked up Muir. He thought Muir's theory of glaciation made sense and encouraged him to publish his views. When Muir discovered the first "live" glacier on Merced Peak the next year, it corroborated his theory about the work of ice.

Glaciers accented land forms already present. They chewed out sawtooth peaks on rock phalanxes and scooped out cirques. When glaciers quarried opposite sides of the same ridge, they carved a serrated edge, creating a line of arêtes. If cirques formed on three sides of a peak, they chiseled a matterhorn as in Matterhorn Peak on the northeast rim of the park. Before glaciation, the Sierra Crest that marks Yosemite's eastern border was neither particularly rugged nor dramatic. After ice chiseled it, the Sierra Crest was spectacular.

* * *

A glacial erratic, markedly different from the country rock around it, was dropped here by the last glacier on Taft Point.

The shearing off of panels of granite parallel to the surface creates uneven hollows and humps in the cliffs below Cloud's Rest.

Pywiack Dome rises just north of Lake Tenaya and, like most of the other domes in Yosemite, was formed by the exfoliation of slabs of rock that left a rounded pate of granite 8,851 feet high.

Intersecting vertical fractures created the twin towers of the Cathedral Spires. Looking west from Sentinel Dome, they rise dramatically some 2,000 feet above the valley floor.

TOPOGRAPHY OF YOSEMITE

Yosemite Valley, created by the Merced River and the effects of glaciation, clefts the rugged terrain of Yosemite National Park on a nearly east-west path. Almost parallel to the main valley runs the Tuolumne River, through its "Grand Canyon" that ends in the dead, dammed waters of Hetch Hetchy Reservoir. Huge granite domes and spires, the result of massive exfoliation, flank both valleys and evoke the spirit of Yosemite. Many domes rise out of and high above the forested slopes. The resulting contrast between somber green and pale, gleaming granites is the quintessential trademark of Yosemite.

The surface plane of Yosemite rises from around 2,100 feet near the Arch Rock entrance on the west to over 13,000 feet at the crest of the Sierra Nevada on the east. Most peaks rise to 11,000 or 12,000 feet; Mounts Lyell and Maclure, each flanked by its own living glacier, rise over 13,000 feet. At the crest, the Sierra Nevada drops off sharply to the valley below, the result of a massive fault system that lifted the Sierra Nevada mountains as a block. The complexities of the terrain create a map with countless squiggling

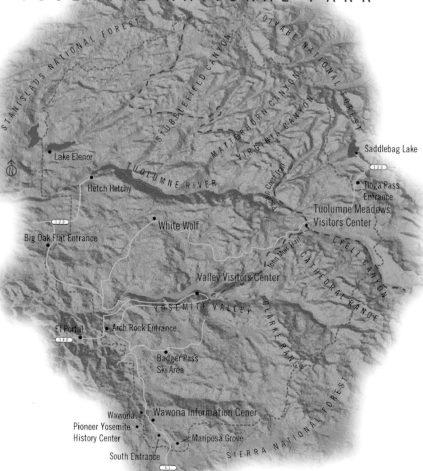

YOSEMITE NATIONAL PARK

STANISLAUS NATIONAL FOREST

TOIYABE NATIONAL FOREST

STUBBLEFIELD CANYON

MATTERHORN CANYON

VIRGINIA CANYON

Lake Elenor

Saddlebag Lake

TUOLUMNE RIVER

Hetch Hetchy

Pacific Crest Trail

Tioga Pass Entrance

Tuolumne Meadows Visitors Center

White Wolf

Big Oak Flat Entrance

LYELL CANYON

John Muir Trail

CATHEDRAL RANGE

Valley Visitors Center

YOSEMITE VALLEY

CLARK RANGE

El Portal

Arch Rock Entrance

Badger Pass Ski Area

SIERRA NATIONAL FOREST

Wawona

Wawona Information Cener

Pioneer Yosemite History Center

Mariposa Grove

South Entrance

Divide Pea

Badger Pa
Ski Area

Sing Peak

Sierra National Forest

Buena Vista Crest

Wawona Dome

Wawona

Mariposa Grove

Fish Camp

Note: Although it may look like a photo-graph, this image is actually a computerized, extruded, topographic view. It was created using digital elevation models derived from the United States Geological Survey (USGS) satellite maps and traditional, flat USGS topographic maps.

To prepare the extruded topo map, data from the USGS was downloaded from the Earth Science Information Center to a personal computer and converted into a three-dimensional model. There, a flat or "birds-eye" version was rendered which simulates a direct overhead view of the region (the end sheets on both inside covers of this book were reproduced from this version). The flat version was then tilted in order to create a view of the area from an angle 20 degrees off the horizon. Shadows, textures, and colors were added to represent a view that one might see from space.

The climate cooling that spawned glaciers began around 3 million years ago. Tromping wet-footed in the mountains after a cold spring, through late snowbanks surprisingly deep, I get a firsthand inkling of how glacial times arrived—snowflake by snowflake, compacting into crusty patches, coalescing into blankets of ice that lasted later each year and evolved into permanent packs. Ice fields form when the total amount of snow each year exceeds that disappearing through evaporation and melting in the summer. Massive ice fields covered much of the high recesses of Yosemite. At a thickness of around a hundred feet, ice begins to flow outward under its own weight. When it flows downhill, a glacier is born.

Glaciers broadened the V-shaped river valleys into the U-shapes that can be clearly seen from Glacier Point. This resulted in flatter valley floors and steeper sides. Ice also straightened the zigzag stream valleys by shaving off protuberances along the way.

The glacial history of Yosemite Valley is difficult to reconstruct, although it can be said that the earlier glaciers were the most massive, filling the valley like a white porcelain pudding nearly a million or so years ago, covering Glacier Point and lapping the face of Half Dome. Subsequent glaciers were smaller and did little to alter the walls of the valley. The subsequent shaping of Yosemite Valley has resulted from the processes of rock falls, spalling, and exfoliation.

* * *

Ice carries sand that, with the glacier's great weight, smoothes and polishes the rock surfaces over which it rides. Kneeling on Polly Dome, I place one hand on a patch of glacial polish, the other on open granite. The polished area is smooth and gleams in the sunshine, markedly warmer to my palm than the unpolished, granulated surface. The polish that once veneered most of this rock face has been tattered by weathering and is now as perforated with holes as a slice of swiss cheese.

Geologists reconstruct glacial extent by these icy signatures: glacial polish, glacial striations, chatter marks, erratic boulders, and moraines. Rocks embedded in the ice leave gouges called striations, or they may chip away the bedrock in crescents, leaving scalloped-shaped chatter marks like lots of smiley faces. Glaciers also toted many boulders downhill, and when the ice melted, left the boulders scattered like a careless child leaves its toys. Two-thirds of the way up Lembert Dome, boulders cluster like Henry Moore figures arranged in a sculpture garden. Called "erratics," many such boulders are markedly different in rock type from the local country rock on which they were discarded and indicate transport from a distant source.

Moraines mark the farthest reach of a glacier, piles of dirt and rock and unsorted debris scraped together and dropped when the glacier no longer pushed forward. Several moraines are visible in the park. One parallels the road to Mirror Lake. Another, east of Bridalveil Meadow, formed around 20,000 years ago. Just west of El Capitan, a glacier delayed just long enough during its retreat to leave an even larger "recessional" moraine. As the glacier

Rock faces at the Fissures show patterns of jointing that formed when granite masses expanded, freed from the pressure of rocks above them, and left a network of cracks. Freezing and thawing attacked the cracks, shattering the cliff faces, forming embroidered tapestries of light and shade .

finally melted and no ice remained in the valley, a lake some five miles long filled the depression upstream from the dam formed by the El Capitan moraine. Gradually, sediment carried by the Merced River filled in the lake and produced the valley floor as it is today.

Although Yosemite Valley's present wide, flat-bottomed valley is the result of glacial plastic surgery last occurring some 750,000 years ago, spalling and rock falls have shaped the upper cliffs more than ice. Yosemite Valley is a beautifully unique combination of rock type, erosion, and a little glacial hewing.

* * *

Striping any steep slope in Yosemite are bare streaks where no trees grow, chutes made by rockslides charging down the cliff face on a more or less regular basis. Since a warming trend melted Yosemite's glaciers (two tiny ones remain on Mounts Lyell and Maclure), most of the landscape changes have come from rockfalls and stream erosion. Rockfalls flush debris downslope every winter, where it accumulates at the base of avalanche chutes. Occasionally, volumes of rock let loose in stupendous rockfalls, as at the Rockslides below the heavily jointed cliffs on El Capitan's west flank. The original Big Oak Flat Road was built on top of the Rockslides. Later slides destroyed the road, and it has since been replaced by a new route with several tunnels. During the Owens Valley earthquake of 1872, Muir and others watched enormous rockfalls cascade off the walls. The avalanche of rock that lies below Slide Mountain on the eastern edge of the park was so huge that it snatched the whole mountainside downhill and blew it across Piute Creek, where it slithered up the canyon wall on the other side. In the winter of 1982, a rockfall cascaded onto the road two miles east of the Arch Rock Entrance Station and had to be blasted out to reopen the road. The newly exposed granite faces remain an astonishing, blazing white.

* * *

Waterfalls may be simply water plus air plus gravity, but there is something magical about the crisp, lacy scrims of spray charged with air that hover like great white moths against the granite walls of Yosemite. The park has four of the ten highest waterfalls in the world. Vernal Fall drops 317 feet, Nevada Fall, 594 feet, Bridalveil, 620 feet, and Yosemite (a total of three falls), 2,425 feet.

Waterfalls form on "hanging" tributary valleys, carved by tributary streams that had much less erosive power than main streams. They could not cut as rapidly or deeply as the larger waterways and over time came to enter the main valley above its base level, which had been excavated by the more powerful streams. Much the same thing happened when ice filled the river valleys. The main ice flow, thicker and much more erosive than its tributaries, cut canyons more deeply than the smaller glaciers could. When the ice retreated, it left steepened cliffs and tributary streams that debouched into space far above the floor of the valley. The erosive impotence of these tributaries shows at the brinks of the falls: water has been able to carve

GLACIAL POLISH

Glacial polish forms as ice drags imbedded sand across the surface of the bedrock, smoothing and polishing it until it shines. Striations are formed from rocks and larger debris carried by the ice that leave long grooves in the direction of the glacier's travel. Glaciers also leave their calling cards in erratic boulders and in the asymmetrical molding of domes that are reformed by the sliding ice cover. Moraines snake across the valley, low hills made up of unsorted debris a glacier dumped when it began its final retreat. The U-shaped valleys are the glacier's large-scale work made out of former steep and narrow V-shaped river valleys.

only a shallow nick in the last thirteen thousand years or so that the tributary canyons have been free from ice.

All Yosemite's falls are at their most spectacular in April, May, and June, when three-quarters of the snowpack melts. In summer and autumn, Yosemite Falls often dries up, and Bridalveil becomes a wispy shadow of its former self. But come spring, mists boiling off in clouds obscure the base of the falls—a billowing water world on the fly, rainbows resonating in the spray.

* * *

Another Yosemite trademark is its magnificent climbing walls. The strength of Yosemite's granite plutons makes it one of the prime mountain-climbing areas in the world. As soon as the park was discovered, peak after peak was climbed, including the first ascent of Mount Lyell, Yosemite's highest point, in 1871. Half Dome was scaled in 1875.

The face of El Capitan remained unscaled until 1958. At thirty-six hundred feet high, it is one of the loftiest unbroken faces in the world. The rosy beige granite of El Capitan is highly siliceous—that is, it has a high content of quartzlike minerals, which makes for unusually strong, well-cemented rock. So solid is El Capitan's granite that it resisted glacial erosion and has weathered back relatively little since glaciers left the valley.

The first climb of El Capitan was completed on November 11, 1958, by Dick Calderwood, Warren Harding, George Whitmore, and Wayne Merry. They had only very primitive equipment, and the exertion, determination, and discipline required of them in pioneering a route over a period of two months was Herculean. They had to go up each day and come back down, going a little higher each time, since the Park Service would not allow them to sleep on the face. Merry remembers the terrifying dehydration they suffered but also the glory, after the final all-night push, of "coming over the rim into the first sunlight."

Since then, El Capitan has taken many lives and threatened others and remains a dangerous climb. But the splendor of the ascent, the almost overwhelming challenge, will continue to entice climbers, even though the uniqueness of the first climb, like Bunnell's first view of the valley, can never be duplicated.

Crustose map lichen, one of the most plentiful and common lichens in Yosemite, frequently encrusts granite with brilliant chartreuse patches. Lichens also grow on soil and bark, in many colors from black and brown to mustard yellow and gray green.

FOLLOWING PAGES:

Bridalveil Fall flows in the fullness of a wet spring, sending up mists of spray that blend with the vapors of a late spring storm.

As eternal and impervious as granite appears, rocks embedded in glacial ice may chip away its surface in crescents, leaving chatter marks. Over centuries, the work of a cobble combined with the action of flowing water can scoop out potholes, and soil blown into cracks can provide nourishment for plant seeds.

Yosemite National Park ranges from two thousand feet in the lowest valleys to more than thirteen thousand feet at the summit of Mount Lyell. This altitudinal range provides a great diversity of habitats for plants and animals, ranging from the richly varied conifer groves of the park's lower and midaltitudes to the distinctive vegetation of subalpine meadows and forests to alpine retreats, endowing Yosemite with some fourteen hundred species of trees, shrubs, and herbs.

It takes a while to get acquainted, but soon one can recognize the individual characters of the trees: the giddy incense cedars, the tiered white firs, the feathery sugar pines with their opulent cones, the long, glossy-needled ponderosas and Jeffrey pines with their jigsaw-puzzle bark, the winsome willows, the flouncy whitebark pines.

* * *

In the valley, the oaks cast a flickering shade with their lobed shiny leaves that in the fall turn butterscotch and crimson. When the Ahwahneechee roamed the valley, they depended upon black oak trees for acorns. A family needed as much as five hundred pounds a year. To ensure big acorn crops, they frequently set fire to the meadows. Fire cleans out the saplings of competing trees like incense cedars and white pine, gives the oaks more sunlight, and does not injure mature trees. The vigorous sprouting of black oaks from the base since the big 1990 burns

A sentinel Jeffrey pine fringes the view of El Capitan across Yosemite Valley and catches the morning sun in its glistening branches.

TRIUMPH OF TREES
PLANT LIFE OF WOOD, SLOPE, AND MEADOW

shows their positive response to fire. Beginning with the Fifth Cavalry in 1891 up until very recently, federal policy was to fight all fires as soon as they were spotted. As a result, perhaps 90 percent of the pure black oak groves in Yosemite Valley are gone. Today, only around 143 acres remain.

Growing along with black oaks in the valley are sturdy, long-needled pon-derosa pines that used to grow impressively large in the open, cathedral-like colonnades that Muir memorialized. Their distinctive bark scales off in thin, jigsaw-puzzle pieces in a rich panoply of colors—purply brown to salmon pink, mauve to tan, old rose and madder, often emphasized by pads of brilliant green moss growing up the bole. With their extensive root sys-tems, ponderosas resist drought better than other evergreens on the Sierran western slope. Sometimes trees a hand's breadth tall will have a root system two feet long, and a twelve-inch sapling can reach water five feet underground. This adaptation served the ponderosas well during "monster droughts," recorded in tree stumps and trunks submerged and preserved in local lakes, that lasted between A.D. 891 and 1112 and between 1209 and 1350. These droughts lasted longer and were more severe than any the Sierra Nevada has had in nine thousand years. Should they come again, they would decimate plants, animals, and two-thirds of California's water supply.

Oak and ponderosa seedlings need a lot of light, and the invasion of small trees and underbrush—the result of fire suppression policies—have impacted Sierran forests, replac-ing open floors in the woods with chock-a-block undergrowth. Shade-tolerant trees like incense cedar ("incense" for the penetrating aroma of its leaves when crushed) and white fir have increased, to the detriment of the ponderosa and oaks. Incense cedars, in fact, may be one of the most successful conifers in the central mountains. They are striking and graceful trees, fast-growing, with distinctive fanlike branches that tilt every which way. Their thick, leafy branches effectively shade out oak and pine seedlings.

White firs also increased as long as fire suppression was park policy. The trees are vulnerable to fire because of their thin bark and because their whorled branches reach clear to the ground. Fire-resistant trees generally lose their lower branches, preventing flames from sweeping up into the crown and killing the tree. Fire easily climbs white fir, turning it into a torch.

* * *

The entrance at Wawona (the Indian name for the Big Trees) or at Big Oak Flat at 4,575 feet leads up into the midelevation coniferous forests that occupy a band between six and eight thousand feet. These are the forests primeval, tapestries woven in emeralds and dusky browns, repeated for miles. At this altitude, the trees are preponderantly conifers. Winters are cold here, and this altitude catches the greatest yearly precipitation on the western slope of the Sierra Nevada—between forty to sixty inches, 90 percent of it as snow. By July,

Dogwood, which grows along stream edges, lights the woods in early spring. The riparian zone of Yosemite supports a rich green band of sedges, willows, alders, California bay trees, broadleaf maples, and cottonwoods, all trees that need ample water during the growing season. In the spring, when these trees leaf out, the tender greens gladden the streamside, and the blooms of dogwood light the dark woods like petaled candles.

drought sets in, and precipitation drops to half an inch, often less in August. Waxy-coated needles, with their small surface area, lose much less water than the thin, flat leaves of deciduous
trees, a distinct advantage in this climate in which summer droughts are common. They can
also photosynthesize on favorable days during winter.

And here at these midaltitudes grow the Paul Bunyan trees of the western slope,
the Big Trees—*Sequoiadendron giganteum*. In late April I stroll the pleasant paved path that
wends through the Mariposa Grove. Snow patches still lie on the ground, and water gushes in
every ditch. With around five hundred mature trees, this is the largest of Yosemite National
Park's three groves.

The first record of the Big Trees appears in the diary of Zenas Leonard, the
"clerk" and recorder on Joseph Walker's cross-Sierra expedition in 1833. Traversing the ridge
north of Yosemite Valley between the Merced and Tuolumne valleys, Leonard was one of a fifty-
member scouting party led by Joseph R. Walker, and he described some "incredibly large" trees
in his diary, which was printed in 1839. The mid-nineteenth century rediscovery of the oversized trees, and the awe and curiosity with which the public reacted to them, eventually resulted
in their protection. The genus was not finalized as *Sequoiadendron* until much later, to differenti-
ate it from the *Sequoia*, the redwood, which it resembles superficially but from which it differs
significantly in cone structure, wood strength, ease of seed shed, and method of reproduction.

Because of their size, Big Trees naturally attracted loggers, and many groves
disappeared soon after their discovery. The wood is brittle and tends to fragment when the
trunk hits the ground, so most of the wood was wasted and ended up in shingles and fence
posts, grape stakes and pencils, and, most ignominious, toothpicks. Logging of Big Trees is now
completely curtailed.

A*lthough Sequoia redwoods may be taller than Big Tree Sequoiadendron, the latter are much more massive. With their thicker branches and vast trunks, a Big Tree may weigh up to two million pounds, among the largest of living things. But as John Muir noted, "So exquisitely harmonious and finely balanced are even the very mightiest of these monarchs . . . there never is anything overgrown or monstrous looking about them."*

Whitebark Pine	**Red Fir**	**Sugar Pine**	**Ponderosa Pine**	**Giant Sequoia** *(Sequoiadendron giganteum)*

300 Feet

250 Feet

200 Feet

150 Feet

100 Feet

50 Feet

American readers were as captivated by the Big Trees' size as by their great age. Ring counts commonly indicate ages of up to twenty-one hundred years, and survival of three thousand years or more is not out of the ordinary. The General Sherman Tree in Sequoia National Park is estimated at thirty-five hundred years, a tree that truly did germinate in Old Testament times. One biologist estimates that General Sherman, 275 feet high and 36 feet at the base, contains more volume than any other tree discovered to date. In 1978, the General dropped a branch almost 7 feet in diameter and 150 feet long, bigger than most trees east of the Mississippi.

Big Trees exist only on the western flank of the Sierra Nevada, in a band 250 miles long and 15 miles wide, and at elevations between three thousand and eighty-nine hundred feet. The seventy-five known groves occupy only thirty-six thousand acres of the more than 19 million acres of the Sierra. The trees grow only on well-drained ridgetops and rolling ground, or in moderate swales that have damp, deep soils—often where snow lies late, and always where abundant sunlight is available. Given their limited range in California, it's surprising that they survive in botanical gardens in a wide range of climates from London to the Black Sea.

Fossils of *Sequoiadendron* indicate that the trees once were numerous in Nevada 2 million years ago, when Nevada's climate had twenty-five to thirty-five inches of annual rainfall. The faulting and elevation of the Sierra Nevada during the last 10 million years caused a rain shadow to develop to the east. The Big Trees were forced to move westward in order to survive, probably reaching the park's eastern boundary 10 million years or so ago. Attrition and warming climate during the last 2 million years have reduced the groves to their present size.

The cones of such huge trees are surprisingly small, armored with a few large, thick, woody scales. The size of hen's eggs, the cones mature at the end of their second summer but remain closed and bright green for twenty or thirty years. Seed production is prodigious, with up to two or three hundred seeds per cone and up to two thousand cones per tree per year; each tree has the potential of yielding half a billion seeds throughout its lifetime. Profligacy is necessary because the odds are that only one tiny seed in a million will sprout, and only one in ten thousand of those will reach old age. The tiny eighth-inch seeds contain so little food reserves that they must make contact with soil immediately, and seedlings must have moisture and abundant sunlight from the moment they sprout. More than half a year's crop doesn't survive the first summer, having been nibbled on by mammals and insects and birds, consumed by fungus from snow burial, or broiled by overheated soil.

Once established, however, Big Trees are speedy growers. Their phenomenal growth rates consistently outstrip those of other native trees in height and girth. In girth, *Sequoiadendron* expands almost three times faster than other trees in a mixed forest. They grow in height 20 percent faster than ponderosa pines, and almost twice as fast as other native

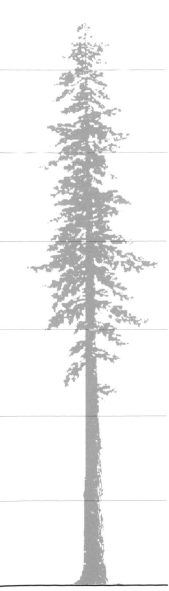

Sequoia Redwood

The giant sequoias, while not the tallest, are the most massive trees in the world, growing to diameters of more than thirty feet—but that gives little sense of their monumentality. As one early writer observed, "What idea of Charlemagne would you get from his tailor's measurements?" .

C alifornia black oak is the most common broadleaf tree found in Yosemite Valley's mixed coniferous forests.

conifers, up to two feet a year in length, a rate that gives them considerable advantage in over-topping other species.

Since Big Trees reproduce only by seeds, the challenge is how to release them from such solidly closed, long-lived cones. In this the trees depend upon crucial help from chickarees (aka Douglas squirrels) and a small long-antennaed beetle, plus an occasional fire to hustle along the drying of the scales. Perhaps because chickarees don't find enough nourishment in the Big Tree's minute seeds, they prefer to eat the scales of two- to five-year-old cones. Peeling off the scales like an artichoke releases the seeds, which may drop to the ground—or if they remain in the cone, they may survive to sprout in a squirrel's midden, refrigerated beneath the snow.

The chickarees' irritable chittering, mewing, clacking, and buzzing accompany anyone who walks through the *Sequoiadendron* groves. Because they live in dense, often dark, forests, chickarees depend upon sound rather than sight for communication. They are incredibly efficient and speedy eaters: one chickaree was clocked at cutting five hundred cones in thirty minutes. In winter, they are active over the snow, bounding easily across it, busy except during the heaviest winter storms.

Long-antennaed beetles lay their eggs on sequoia cones that are at least four years old. The eggs hatch into larvae that tunnel into the cones' vascular system, cutting off their water supply, causing them to dry out, the scales to open, and the seeds to drop. This particular beetle and the sequoia have had a long-term symbiotic association of perhaps 150 million years, whereas the relationship with chickarees may have been in place only a meager 20 million years.

So little *Sequoiadendron* regeneration had taken place since 1900 that studies were instigated to find out why. The investigations documented fire's importance for the Big Trees' healthy growth and reproduction. Frequent natural fires tend to be small, and in the Mariposa Grove they would have come, had they not been suppressed, every dozen or so years. Tannin gives the bark its ruddy color and acts as a preservative, discouraging insect attack as well as insulating the inner growing layer against fire. Without the flammable resins of most conifers, mature trees rarely succumb to fire. Fire reduces competition for water and nutrients by removing brush and other trees, allowing sunlight in, adding nourishing minerals to the soil, often destroying pernicious fungi, increasing the soil's ability to hold water, and providing fiery updrafts that dry the Big Tree cones and open their scales.

Since the 1960s, the necessity of fire in forests has been documented and recognized, but policies have been slow to change. The beloved image of Smokey Bear did a lot to remind people to douse their campfires, but it also instilled in the American mind the idea that all fires are bad. Under firefighting policies that lasted into the 1970s, a thick understory developed in the sequoia groves, and white fir and incense cedar invaded, shading out any

Titanic giant sequoias dwarf white firs in the Mariposa Grove. The largest, the Grizzly Giant, has leaned off-vertical for over a century and suffered many lightning strikes, impressing Professor Joseph LeConte in 1870 with its "great life, decaying but still strong and self-reliant."

Sequoiadendron seedlings that might have sprouted. In the 1970s, the idea of "prescribed burns" debuted as a management tool when the park realized that the accumulation of brush and debris had the potential to produce bigger, hotter, and more disastrous fires. In Yosemite, more than twenty-eight thousand acres have been carefully burned, replicating the small "cool" burns that traditionally swept through the Sierran ecosystems every ten or fifteen years. The more severe the fire, the more *Sequoiadendron* seedlings establish, and even adults show a burst of growth after a blaze.

Seldom do Big Trees die of old age. Generally death comes from the gradual weakening of the tree. Their shallow root system leaves them very vulnerable to wind throw, especially since their crowns are usually higher than the forest canopy and beset by greater wind torque. After a Big Tree falls, it may take a thousand years to decay, slowly leaking nutrients back into the soil.

A fence surrounds the Mariposa Grove's famous Grizzly Giant. With a 30.7-foot diameter at ground level, it rises some 300 feet and is estimated to be twenty-seven hundred years old. Many of its upper branches are broken off, making the remaining ones look even bigger and more muscular, like the arms of some old prizefighter. Cirrus clouds swim by high above it, and for a moment I have the strange sensation that the tree moves and the sky stands still. Other trees in the grove sway, their branches bow. The Giant remains unmoving, immutable, beyond all answering.

* * *

Just pushing out of the ground in the Mariposa Grove in late April are raspberry pink snow plants, buds tightly packed, leaf edges fringed with white, looking like small rockets poised for take-off. Not until July do they bloom an outlandish, shocking pink, often spotlighted dramatically, one here, one there, small torches sizzling in a shaft of sunlight. When I detach a single blossom I find its outer bracts and stem covered with infinitesimal, transparent glandular hairs that render the plant sticky. The small bell-like flowers have ten yellow stamens that leave a line of pollen against the tube and an astonishing salmon pink pistil in the center. They are saprophytes, without chlorophyll, unable to make food on their own, and dependent upon their giant-boled munificent hosts. They usurp little of the energy production of a ponderous ponderosa or sugar pine, charitable trees who spare a dime to these freeloaders.

* * *

Via Tioga Pass, at 9,945 feet, I enter Tuolumne Meadows, the best known of Yosemite's subalpine meadows and the largest, roughly ten by fifteen miles in size. In the past millennia, the meadow was forested, but trees retreated when the meadow soils held too much water, dammed by underground granite ledges that block the Tuolumne River at the Tuolumne Cascades. Myriad wildflowers—shooting stars, corn lilies, yellow ragworts, larkspur—bloom against a background of silken sedges and rushes, the nourishing plants that made these

Dogwood flourishes along the Merced River.

FOLLOWING PAGES:

At high spring flow, the Merced River catches morning reflections.

meadows so popular with sheepmen.

Lodgepole pines, trees typical of the subalpine, wreath Tuolumne Meadow. Lodgepoles grow to altitudes of ten thousand feet, often intermixed with mountain hemlock and, at the upper edge of their growth, with alpine whitebark pine. In contrast to the richly mixed conifer forests of lower altitudes, both lodgepole and whitebark pine often grow in single-species stands.

I shelter in a cluster of lodgepole pines, short sturdy trees with needles growing in bundles of two. Nearly every branch includes a clump of new male cones at the tip plus female cones in all stages of development, from half-inch maroon nuggets shaggy with little spines to two-inch-long purplish-brown cones, tightly closed. The trees clearly are in good health and getting about the business of reproduction with admirable efficiency. The branches curve into Cs and Js and Ss, round cheerful consonants curling up at the tip in the most optimistic fashion.

Tall lodgepole pines were used as tepee poles, hence their common name. In the Rocky Mountains, lodgepole cones remain closed on the tree and do not open until heated. In the Sierra, some of the cones have a waxy coating that keeps seeds contained. Others are not coated and can open and regenerate without fire, although fire insures a much more prolific seed fall. Chipmunks and chickarees eat lodgepole seeds, as do several birds like crossbills and Clark's nutcrackers, the raucous jays of high altitudes. Nutcrackers extract the seeds and then hide them in the ground, and those they don't find may germinate the following spring.

Often growing with lodgepoles are mountain hemlocks, distinctive, graceful trees with languid, nodding spires. At the lower edge of their range, mountain hemlocks grow where snow lasts late, often on north-facing slopes. Much of the influx of trees fringing Tuolumne Meadows came after livestock grazing ended, when the ground had been broken and disturbed by hooves, thus providing an ideal situation for lodgepole seeds to sprout. In 1870, Joseph LeConte noted grazing flocks of three to four thousand sheep each, herds totaling twelve to fifteen thousand sheep a summer. The meadows suffered accordingly.

Muir's distress at overgrazing here largely motivated his campaign to establish a national park. In 1889, Muir camped in Tuolumne Meadows with Robert Underwood Johnson, who was associated with *Century*, the most influential magazine of the time. Johnson saw first-hand the damage that Muir's "hooved locusts" did when they trailed through the meadow, and he offered to publish anything Muir wrote on the subject. Muir's articles were undoubtedly instrumental in persuading Congress to sequester 757,617 acres for Yosemite National Park in 1890—not including Yosemite Valley and the Mariposa Grove, which remained in state hands until 1906.

* * *

The subalpine zone frays and fritters out at its upper edges where it encounters the rigorous alpine conditions of rocky, wind-exposed ledges, aridity, and high solar radiation. Trees at timberline are most likely lodgepole or whitebark pine, and even these must cluster in pockets of protection and warmth. Winter may last ten months out of the year up here, and frost may whiten the ground any night.

The trademark tree of high altitude is whitebark pine, which may become established up to twelve thousand feet, although it is more common between eight and ten thousand feet. Whitebark branches interweave to form petticoats of greenery above which spires rise, either flagged or with stalks scoured bare by the ice-laden winds of winter. Frequently stunted at this altitude, a six-inch-diameter trunk may be fifty years old, as John Muir, counting rings, discovered.

Lavender in color, very sticky and resinous, whitebark pinecones remain snugly closed until maturity. The seeds are wingless, which would condemn them to dropping near their parents were it not for Clark's nutcrackers' penchant for hiding the seeds of whitebark pines. Chickarees, deer mice, and chipmunks also harvest seeds fallen from open cones, although deer mice, with their sharp sense of smell, probably find and eat most of the seeds they've hidden.

* * *

Because of the severity of the alpine environment, plant genera here have affinities to plants growing at lower altitudes under similar circumstances, such as those in arid California or the neighboring Great Basin desert. With so little water and high, desiccating winds during the growing season, plant growth is patchy. Many plants are grayed with hairs to protect against evaporation and UV radiation. The richest alpine communities in Yosemite occur on unglaciated plateaus, the "nunataks" unpared by ice, such as the Dana Plateau, that acted as refugia for plants that in other places were destroyed by ice cover.

Granitic gravels on steep slopes drain quickly. They are almost impervious to weathering into soil. After snow melts, there is very little moisture during the summer except for an occasional fleeting storm. Only below late-lying snowbanks, where moisture drizzles downward from snowmelt, is there a rich assortment of sedges and wildflowers.

Alpine plant communities are usually so small that one can literally stand with each foot in a different community, put a hand in a third, and have a fourth within view. But what alpine plants lack in luxuriance they more than make up for in charm, their brilliant, lilliputian flowers displayed against a view as big as eternity.

THE FORESTS OF YOSEMITE

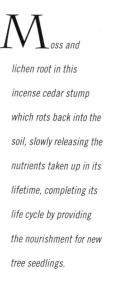

The rich and varied tree species in Yosemite tend to grow in altitudinal bands or trail along streamsides. Bunnell remarked in 1851 that "the valley at the time of discovery presented the appearance of a well-kept park. . . . There was then but little undergrowth in the park-like valley, and a half day's work in lopping off branches. . . .enabled us to speed our horses uninterrupted through the groves." The groves he refers to were those of black oaks, with their distinctive spreading and gnarled branches, and the columnar ponderosa pines. Both grow in the lower altitudes of the park around 4,000 feet, in the past kept open by fires which cleared out the competing underbrush but left the oaks and pines healthy.

Going up in altitude, the forests contain a richer mixture of conifers, including incense cedar and white fir, easily told by their distinctive leaves. Cedar branches end in splayed fans covered with scalelike leaves. White firs have single, often curved needles two inches long that separate them from pines with their

Moss and lichen root in this incense cedar stump which rots back into the soil, slowly releasing the nutrients taken up in its lifetime, completing its life cycle by providing the nourishment for new tree seedlings.

Giant Sequoia and Cone

White Fir (with moss-covered giant sequoia trunk)

Red Incense Cedar

Western Juniper

Red Fir

Red Fir Cone

Sitting on top of Yosemite Falls in early spring, I watch a western fence lizard snuffle about, not quite up to running speed on a cool day, quite different from the three that I saw skittering across the crest of North Dome late last summer. They match the granite, their charcoal-gray backs speckled with darker gray chevrons. Having only five choices simplifies lizard identification in Yosemite: there are only sagebrush, two whiptails, an alligator, and a western fence lizard, the only Yosemite lizard that has a blue belly and throat. Like most other lizards of Yosemite, they occupy a wide altitudinal range, prowling boulders and ledges where they catch spiders and insects.

It strikes me how few of Yosemite's animals I've seen, even though I've been here at various times of year. Not counting insects, I'd guess around twenty. Of the animals of Yosemite—roughly 30 species of reptiles and amphibians, 240 species of birds, and nearly 80 of mammals—it's unlikely that many people have seen them all. The best places to see the greatest diversity and number of animals is in the mixed coniferous forests, which are rich in insects, birds, and small mammals, because they offer a variety of niches. But even then, it is not likely a visitor will see a mountain lion or a nocturnal flying squirrel or a shrew. Bears? Maybe. Chickarees and Belding's ground squirrels, marmots, juncos, dippers, and Clark's

With good reason Bunnell wrote that "comparison of the Yosemite Falls with those known in other parts of the world will show that in elements of picturesque beauty, height, volume, color and majestic surroundings, the Yosemite has no rival upon earth."

LIFE ON WING AND WATER
WILDLIFE OF THE VALLEY

nutcrackers? Probably, if you go where they are. Insects? Absolutely.

Creatures that crawl or bound on four legs, prowl on eight legs, slither along on their bellies, or twinkle along on six legs abound in Yosemite. Six-legged insects, by far the commonest of land animals because of numbers—possibly ten thousand species or more inhabit Yosemite—provide endless hours of fascination. Some are more likely than others to be seen by an observant hiker. Drape a wet poncho over some branches, and it's soon full of tiny winged creatures of various persuasions, shapes, and behaviors. Check a boulder or a tree trunk: there's probably a troop of ants on various errands. Sit quietly and watch a flower-rich meadow: on a sunny day there will be dozens of bees, wasps, flies, and butterflies harvesting its bounty of nectar and pollen, assuring next summer's bloom. Peer into a quiet pool: marvel at the underwater inhabitants pillaging their dinner-plate-sized surroundings.

Remember: Insects are not the smallest creatures by far in Yosemite. There are the microscopic billions who live between soil particles, the myriads who occupy every thimbleful of water, like *Giardia lamblia* that now infest every Yosemite stream. If you become infected, "Giardiasis" is what your doctor will call it. Your name for it may be unprintable when the organism becomes encysted in your very own intestines—the reason why all water in Yosemite must be treated by boiling, filtering, or adding purifiers.

* * *

I am intrigued by the tiny things that populate the earth, probably an outcome of my myopia, finding the world close to my feet always in focus. It's impossible for anyone to walk in Yosemite and not see ants—crossing the cement walk at the Visitor's Center or scurrying about on an alpine boulder. I have to stop and watch. In my everyday world, there usually isn't time to watch ants react to weather and find their way home, but here there is, and their peregrinations and behavior patterns can entertain a good observer for a long while. Ants communicate by pheromones, chemicals so subtle that a few molecules in the wind can alert another ant to danger or to which path leads to food and how much. When two ants rub antennae, they are communicating fairly complicated coded information to each other through pheromones. They live in a chemical world, not a visual or aural one as we do.

I can also spend long minutes, nose to the surface of a small pool, watching the water beetles zip back and forth between bottom and surface, where they replenish the air supply they carry under their elytra, stiff shell-like panels that cover gossamer wings beneath. Hundreds of species of beetles stalk the park, from tiny pine bark borers and flower beetles wiggling away among the stamens of a buttercup to the long-horned beetles that tunnel into Big Tree cone scales. Beetles are wobbly fliers at best and can often be recognized in the air by their lurching, insecure flight. The exceptions are tiger beetles, often brilliantly colored, yellow and black or iridescent green, which are swift to run and quick to fly on their successful six-legged passage through life.

Half Dome stands sentinel over a valley filled with a diverse wildlife.

Corn lilies grow
below 11,000 feet in
thick stands wherever
soils are kept damp by a
high water table, often
growing head-high.
Also called false helle-
bore, they produce big
parallel-veined leaves
and dense panicles of
small, white, six-petaled
flowers.

Waiting for water to boil at ten thousand feet, I watch the neighborhood assassin flies. They have the alert, intense aspect of efficient hunters, long legs sturdily splayed out in front for a short-field take-off, looking as lethal as F-16s. They are nonbiting flies, not interested in humans but only in smaller insects. From its observation perch, one assassin fly leaps out when a tiny fly wafts by, catches it in spiny legs, inserts its proboscis, and sucks the victim dry. Then it returns to the same perch to watch and wait again.

* * *

Hiking up Rock Island Pass, I top out into a high, busy, buzzy, blazing meadow with a brilliant array of wildflowers. Lovely flowery aromas swirl up into the sunshine. Fuzzy little bee flies, the teddy bears of the Diptera family, butterflies, and tiny day-flying moths service the meadow in a frenzy of nectaring, transferring pollen on basketed legs, going about their business of making next summer. Scores of half-inch, iridescent moths flutter low to the ground, joined by tiny, pale yellow butterflies.

This summer brings out busy clouds of small blues and fritillaries, sulphurs, and Parnassian butterflies. Their small size and large numbers animate the whole profile of the swale so that the meadow appears to move, shift, and shimmer. A bright yellow butterfly flutters erratically, alighting on a wine-red penstemon, each intensifying the other's color. Intermittently, the meadow erupts with butterflies like windborne scraps of bright tissue paper, choreographed to a Mozart quartet. There's a moment's quiet as they feed, then as if by signal they all leap up, twirl, and flutter into the air.

A bee fly alights on an open daisy face. A little round furry bundle with wings held straight out to the side and a long proboscis it could practically pole vault with, it settles in to feed. It does not sting but simply uses its proboscis like a straw to draw up nectar. Flies belong to the Diptera family, the "two-winged" ones. Their second pair of wings is replaced by "halteres," small posts mounted right behind the wings that vibrate at extremely high rates and provide the stability of four wings. A fabulously successful family, flies occupy every available habitat, from brine pools to alpine fells, and use every kind of food, from fresh manure to pollen to vertebrate blood.

In a damp spot, several bees work the plants called little pink elephants. The configuration and weight of this particular species of bee are precisely adapted to the flower's contour. Other bees work the penstemons, irregular tubular flowers in the Figwort family. In a bitter winter, if bees do not survive hibernation, there may not be enough of them to do the necessary pollinating, and the following summer's bloom will be diminished.

One morning as I strike the tent, a dozen minute moths that have no common names, moths no one studies much or knows much about, cling to the damp nylon. There are so many of them, little tiny motes with scaled wings and feathered antennae, living out minuscule lives in a vast breadth of earth and sky, small snippets hatched from pinpoint

One of the larger predators in the park, the resilient coyote, alert and curious, pauses in a spring snowstorm before continuing his probe for a tasty meadow vole or mouse.

eggs, going about their tiny business.

The needle-miner moth, a small moth with black-speckled white wings, lays her eggs at the base of a lodgepole pine needle. They hatch in two weeks, then proceed to mine the needle and spend their first winter inside, mining two more needles the next summer. They pupate in June, and adults emerge in July and August in, for heaven's sake, only the odd-numbered years. In single-species stands, a severe infestation of needle-miner moths can destroy acres of lodgepole leaving Yosemite with "ghost forests" of bare, gray trees.

* * *

The morning that I climb to Inspiration Point, I marvel at the massive numbers of little red berries produced by the manzanita. On closer examination, they are not berries, but galls that infest leaf edges. My eye is also caught by little red balls on the willow leaves, galls made by the tiniest of flies. Galls are fascinating studies in how insects tenant every possible niche in nature, often bending plants to their needs. Yosemite oaks host a whole series of galls, some of them the size of pencil erasers, some pinheads, some spiky and spiny, some like miniature pincushions, some as large as a marble (or larger), some little green or red pimples on the edge of a leaf. Each species of insect forms a distinct and singular gall, an arrangement that's been going on for millennia. Most galls do only cosmetic damage, and the relationship seems to be benign.

Those puzzling "pinecones" at the branch tips of willow branches—the resemblance is startling when they dry to brown—are galls made by minute, nonbiting flies called midges. The female injects an egg into the forming bud at the end of a twig, along with a growth hormone. The egg hatches into a larva that plugs into the plant's vascular system. The plant responds by ensheathing the midge with a characteristic covering, the gall, and shunting some of its food to the larva's use. The larva enjoys protection and an unending food supply until it's time to pupate and change into an adult midge, at which time it bores a pinpoint hole and exits.

* * *

Even the swiftest flying insect is sometimes nailed by a quick-tongued amphibian. In any wet subalpine meadow, pocket pools may hold dozens of little black, vigorously wiggling tadpoles of Pacific tree frogs or Yosemite toads. They spawn from January to early summer at lower, and later at higher, elevations. Pacific tree frogs are the smallest of California amphibians and have an extremely wide range, from Baja California to British Columbia, from low altitude up to the alpine zone. Not only the black stripe through the eye but also the adhesive discs on its toes mark it as a tree frog—a misnomer, since it spends more time on the ground. Tree frogs sing throughout the summer, not just during mating season as other frogs and toads do. Muir welcomed their singing when he channeled a bit of Yosemite Creek through his cabin: "What a cheery, hearty set they are, and how bravely their krink and tronk concerts enliven the rock wilderness!"

A member of the Daisy family, lavender asters grow in dense drifts in Yosemite's meadows.

One population of the Mount Lyell salamander lives on the summit of Half
Dome at 8,842 feet. Some live around Yosemite Falls, others on Mount Lyell, where they were
first discovered in 1915 by University of California zoologists who were running trap lines to
sample the residents of different habitats. Mount Lyell salamanders are endemic to the Sierra
Nevada. Populations are found in scattered localities and may be relics of past glaciations,
which left their colonies small and widely separated, like the Big Trees.

Salamanders seek shade and nestle in crevices and crannies, so casual hikers are
not likely to see them. Lungless, they breathe through their thin, delicate skin, so they must lurk
under logs and rocks to keep moist. They are well camouflaged, too; their skin is dark and
flecked and matches the specific granite where they live. Mount Lyell's have a flatter head and
body than most salamanders and a short tail. They pad about on blunt, webbed toes that adhere
to steeply inclined surfaces and are particularly effective on glacial polish. They have the curious
and distinctive habit of curling their tail forward. They also use it as a stabilizer and a pusher,
pressing it against the rock as an aid in walking.

* * *

At the edge of a cone-shaped pile of dirt a Belding ground squirrel, sleek of
coat, sits up on its haunches and whistles about somebody new in the neighborhood, then bob-
bles, lickety-split, across the flat alongside the creek. It's a great color match on bare sandy
spots and no match at all against the green sedges. The proximity of its holes and the speed of
its transit must be the only things that save it from being hawk food.

Belding ground squirrels burrow in soft meadow soils and leave circular heaps
of dirt that mark burrow entrances. When a squirrel spies an interloper, it runs toward its burrow, stands up on its hind feet for a better look around, and whistles a warning. The ground
squirrels' digging mixes seeds into the soil and aerates it, and their spading in of plant parts and
the addition of their own excrement enhances soil fertility. A dense population of ground
squirrels moves tons of topsoil per acre per year—important, since most of Yosemite is above
the range of earthworms, who accomplish this so efficiently at lower altitudes.

Belding ground squirrels spend most of their year in hibernation, which
requires that they eat almost continually from the time they emerge in late spring until their
retirement in early autumn. During this time they increase their body fat fifteenfold and double
their body weight. Mainly vegetarians, they stuff down flower heads and all kinds of seeds.
Males have also been known to kill the helpless, hairless young from other nests and have been
caught robbing eggs from the ground nests of white-crowned sparrows. You can watch their
sybaritic approach to harvesting meadow grasses: they lie on their backs and pull the seed
heads down with their paws, wallowing in summer's plenty.

Being so numerous, they attract many predators. Several hawks and Clark's
nutcrackers strike from the sky. A weasel is slim enough to wiggle into their holes. A badger

M*iner's lettuce,
with round, fleshy
leaves that surround
the stem, was so-called
because it furnished the
first salads to miners
starved for greenery in
the spring.*

*FOLLOWING PAGES:
A great blue heron
hunts along the edge of
the Merced River.*

ANIMAL COMMUNITIES—
THE DIVERSITY OF LIFE

The wide range of altitude in Yosemite National Park, from around 2,000 feet to over 13,000 feet, engenders a wide range of possible niches for animals to exploit. From the tiniest gnat to the heavy-shouldered black bear, animals frequent the areas in which they most readily find their preferred food. Of the roughly 350 animals in the park, which include mammals, birds, lizards, and amphibians, the casual walker will see relatively few (except perhaps for the ubiquitous ants and flies) in the heavily touristed areas. Most can be seen by knowing where to look and sitting still, quietly aware that you are both watcher and watched.

The most numerous inhabitants of the park are insects, whirring and buzzing, crawling and hopping, nectaring and fluttering. They are found at all altitudes, in all habitats, including water.

The smaller the mammals and the lower they are on the food chain, the more numerous they are. Myriad small furry creatures scamper through the park: chipmunks and voles, mice and ground squirrels and tree squirrels, all vegetarians. The exception is the shrew with a voracious appetite that supports its furious metabolism rate. At high altitude, a distant cousin of the rabbit, the pika, trims stems and dries them to form a hay pile upon which it can feed all winter. Its

A young Belding's ground squirrel sits up on its haunches in a typical ground squirrel pose, checking out the world about it and inviting attack from hawks that ply the sky above.

Bobcat

Red-winged Blackbird

Clark's Nutcracker

Pika

The birds of Yosemite cling to their preferred habitats: red-winged blackbirds yodeling alongside a midaltitude stream, a dipper bobbing along the bottom of an alpine freshet, picking up insects off the bottom as the water cascades over it. Clark's nutcrackers pick up the seeds of whitebark pine at timberline, while spotted owls frequent the deep forests where they find protection, ample trees, and dead snags for nesting.

Of the several amphibians of Yosemite, the Pacific tree frog frequents almost every altitude, even up into the alpine, favoring wet swales with running rivulets. A tiny frog that lives on the ground ("tree" is a misnomer), it has a big voice that fills the wet meadows with bleats and *baa-aa-aas* on warm spring evenings.

This potpourri of animals thrives on a great variety of habitats: low open meadows, high wet meadows, cliffside niches, gentle slopes, tree branches, and stream bottoms. The richness of Yosemite's vegetation engenders a like richness of animal life that animates the air with hums and whines, howls and barks, croaks and calls, and tweets and trills.

REALM OF THE BLACK BEARS

Bears have been a part of the Yosemite scene from the beginning. Almost as adaptable as coyotes, bears that haven't discovered easy pickin's around campgrounds tend to be elusive, although they have no natural predators (unless one counts a crew of parasites). Female bears usually give birth to two cubs a year, born in winter. They remain with their mother for the year, then join her in sleeping through the next winter (black bears do not sink into the torpor of true hibernation). The following spring she sends them on their way and is ready to breed again. Black bears are omnivorous with a predominantly vegetarian diet, choosing greens and bulbs in early season, then fattening up on acorns and nuts before going down for winter. Today's hiker seldom sees bears—only the signs of their existence: the shredded bark of a bear tree with clear claw prints, scat, or the impression of a paw in the dirt.

sniffs through a colony like a metal detector and digs out hibernating ground squirrels. With strong shoulder muscles and large front claws, badgers are able to see while digging because of a transparent inner eyelid that protects their eyes. Its size and energy needs require the badger to consume at least one ground squirrel (or its equivalent) per day.

* * *

In midsummer, birds are everywhere. The short Yosemite summer offers good feeding for the almost 250 species that frequent the park. Knowing what birds to expect in what habitat simplifies bird identification. If I'm in the alpine tundra, for instance, the rigors of the climate shorten my list: the big gray birds with raucous voices are Clark's nutcrackers. The tiny dark birds that spring up from the grass almost under your feet, calling a repeated, high *"Tea-tea-tea-tea,"* are American pipits, recently discovered breeding in Yosemite. If a dozen small birds forage at the edge of an alpine snowbank, they will be gray-crowned rosy finches. The large, pullet-sized bird waddling along the ground, keeping to the willows, is a ptarmigan. Native to alpine areas (although not to Yosemite), ptarmigan were introduced in 1971 to the Mono Pass area by the California Department of Fish and Game, and they have since spread.

Walking on the bottom of clear mountain creeks, or bobbing up and down when they stand, are dippers, sometimes called water ouzels, Muir's favorite bird. They make their nests of moss close by or over permanent, fast-moving mountain streams. A dipper is the only bird adapted to living in perennial, cascading streams, where it walks the bottom as easily as most birds walk the field. It has dense plumage for insulation, waterproofed with oil from special oil glands. Unique flaps shield its nostrils and keep water out of its eyes. Active all year long, its sweet, trilling call is a special delight on a winter's day.

Walking toward Buckeye Pass through airy, open woods, I start up a flock of Oregon juncos, one of the most common species in the park—small sparrowlike birds, brownish gray with a black hood and white outer tail feathers. They have thick, strong bills as befit seed-eaters. The advantage of banding together in flocks is better visibility, several pairs of eyes being better than one, and their combined sounds carry farther than a single sound. When they're disturbed, they gather in a tree and make a sound like a drumstick lightly tapping on the rim of a snare drum, continuing their quick, irritable *"tack tack tack"* until I'm out of sight.

Streamsides and mixed conifer forests are richest in bird species. A hike out to North Dome is a walk through bird music. In the willows along Rancheria Creek a great twittering pulses and wanes, for many birds prefer these shrubs for nesting and cover. Yellow-rumped warblers forage around the willows, and white-crowned sparrows bob from branch to branch in these shrubs, which also provide nest sites and song perches. These sparrows are beholden to snowfall patterns—not enough snow and the meadow dries out too soon: too much snow with too late a melt and the willows leaf out too slowly, leaving fewer places to nest. White-crowned sparrow populations have been known to drop 60 percent when summer comes late after a long, snowbound winter.

One afternoon, I'm having a snack on a rock above timberline, when a Clark's nutcracker flits close by, a jumpy, nervous character dying to get close enough to grab a raisin I dropped. Nutcrackers are unmistakable birds, jay-sized, pale gray with smart black wings and a tail edged with white. Anyone will see them in subalpine pine forests or higher, collecting seeds from whitebark pines. The nutcracker ingests some of the seeds, but collects most of them in a pouch in its mouth, which is large enough to store up to 150 seeds. Clark's nutcrackers may store more than thirty thousand seeds in caches for later retrieval, and some have been estimated to hide ninety-eight thousand seeds in a prolific seed year. They have amazing memories, finding between 60 and 90 percent of their caches. The nutcracker buries seeds over a wide area, and in so doing helps disseminate whitebark pine seeds where they would otherwise be unable to go.

Birders obsessed with warblers could spend weeks looking for the seven species that inhabit the park. But if they knew where to look, they could find them all in one place: Ahwahnee Meadow. Even though the birds all hunt insects, they coexist by foraging in different microhabitats. MacGillivray's warblers rummage for insects in thick shrubbery near the ground. Hermit warblers poke about the middle and upper branches of ponderosas and incense cedars, while larger, yellow-rumped warblers work the outer branches. Wilson's warblers and yellow warblers hunt in the shrubbery along the Merced River; they avoid competition, as Wilson's warblers rummage in the lower leaves, and yellow warblers feed in the top of the willows or in larger deciduous trees. Nashville warblers search the black oaks and maples at the edge of the meadow, while black-throated gray warblers forage in the canyon oaks.

* * *

Whenever I backpack in Yosemite's backcountry, I think a lot about bears. Big bears. Black bears. A black bear is a formidable chunk of real estate, weighing up to three hundred pounds, and could outrun me easily.

As long as there were few people, there were few encounters with bears. By the 1920s, when humans and bears overlapped in the same space, confrontations not only became more frequent but more dangerous. Uninformed tourists, thinking bears were cute, fed them by hand. Bears are not dumb; they soon associated food with two-legged suppliers and proceeded to destroy tents and demolish cars to get at it.

In addition, the park maintained open dumps, which were an irresistible attraction for bears fattening up for winter, when they'll eat anything they can get their paws on. Concessionaires set up "feeding platforms" to entice bears to a central area. Then they set up spotlights that could be turned on at night—prime bear feeding time—and bussed people to see the entertainment. In 1929, some two thousand visitors a night watched the bears chow down. In that year, eighty-one bear maulings occurred.

The turnaround in public awareness finally came in 1973, when the national park reacted to public prodding and recognized that only scientific research could provide answers to "the bear problem," a problem that needed to be handled in a well-informed, sensible, and nonemotional way. By 1973, out of 2.3 million visitors, there were only sixteen bad encounters with bears, and the park issued no citations to misguided tourists.

One would like to think that the "period of ignorance and tragedy"—when tourists took advantage of bears' natural proclivities in order to treat them like a circus act, and reacted hysterically when bears did what they were genetically programmed to do; when no thorough field data on bears was available; and when what you didn't know did hurt you—is over. The park now fines people who leave food boxes out or neglect to "bear-bag" food. In some areas, the park requires the use of "bear barrels," heavy plastic containers in which to carry food.

* * *

When Yosemite opened, the accepted approach to animals was emotional, dominated by the morality of "bad animals" and "good animals," depending upon their usefulness or attractiveness to the public: bunnies were cute, weasels were bad. Nobody considered that, were it not for weasels, we'd be up to our eyeballs in cottontails. Today's camper will never see a grizzly (extinct here since 1924) and likely not a mountain lion, since they've been systematically hunted. In 1927, for instance, the state lion hunter killed forty-three mountain lions in Yosemite. Wolverine, fox, marten, fisher (a small but energetic carnivore), lynx, and coyote were trapped by park rangers until 1925 and now are almost gone from the park. Without fishers, the porcupine population has exploded because it has no control. The shooting and poisoning of coyotes and wildcats to protect domestic sheep also allowed deer to increase, often overpopulating their habitat. The "bad birds," Cooper's and sharp-shinned hawks, no longer breed in Yosemite Valley as they once did. Nonnative species, like trout and Tule elk, were introduced as a ploy to attract more tourists.

Fortunately, that wildlife management approach no longer holds. Now if trout die out in a high-altitude lake because of their inability to reproduce in the short summer season, they are not replaced. The Tule elk were moved elsewhere. Bighorn sheep that disappeared in 1914 because of loss of habitat to domestic sheep and hunting were reintroduced in 1968. They have done well. There are no longer displays of animals doing "cute" things like feeding at the dumps. Today's visitor, more sophisticated and educated, can only welcome park policies that stress maintaining an environment as undisturbed as possible under today's heavy visitation pressures. Now, when you see an animal, it is a special gift, a fascinating glimpse of an animal behaving naturally in its own habitat—one of the treasures that a national park provides.

MULE DEER

Amule deer buck carries a rack of antlers, indicating that he is a mature male. Antlers begin to grow anew each year in the spring, covered with "velvet" until early fall when they turn hornlike and are then shed late in the season. Browsing in the meadows of Yosemite, mule deer eat grasses and sedges as well as shrubbery and wild fruits. Although they are preyed upon by bobcats, mountain lions, and occasionally coyotes, the mule deer suffer their greatest losses through starvation and disease. Large herds roam the park now. Without their former controls of grizzly bears and wolves, they have the potential to overpopulate an area.

One springtime noon I hop a shuttle bus that delivers passengers to Stop 17, the Mirror Lake Road. A glacial moraine flanks one side of the road, a massive windrow of rock debris ten or so feet high, dumped around fourteen to fifteen thousand years ago by one of Yosemite's glaciers. Tenaya Creek larks along the other.

Different shades of spring green light the woods—beautiful feathery greens, catching spring sunlight. Oak leaves open, tender little puckered things with minute bristles at the tip of each lobe, soft green at the base and rosy at the tip. Bluish-green new shoots tip the Douglas fir. Bay leaves give off a wonderful, pungent scent. Dogwoods reach out saucers of ivory flowers. Broadleaf maples dangle flowers; their new leaves, fresh green, are still crinkled like crumpled tissue paper. Cottonwoods break open acid-green heart-shaped leaves out of varnished buds. This distinctive group of trees traces Yosemite's streams, forming a riparian ribbon of green close enough to the banks that their roots have a continuous supply of water. Trees such as cottonwoods that need not withstand drought can afford to have large, thin leaves.

Starlike mosses with big capsules luxuriate in shaded spots. For such lilliputian plants, mosses have an astonishing variety of textures—asterisks, interwoven mats, imbricated

WALKS ON THE WILD SIDE

ALONG STREAMS, SPRINGS, AND FALLS

and overlaid. When dry, they are hardly noticeable, but when wet, they blaze brilliant emerald green. Most of them bear cup-shaped capsules full of dustlike spores, ready to be scattered by the wind. There are more than forty species of mosses in Yosemite, and very little is known about them, making them a wonderful project for an amateur muscologist.

Rock debris that avalanched from both walls of the canyon dammed Tenaya Creek and formed Mirror Lake. The rockfall lies in Tenaya Valley, at the foot of Half Dome on the south and North Dome on the north. Geologists calculated it at 160 million tons, the largest identified in the park so far. Park managers in the last century attempted to maintain the lake's water level by dredging and building check dams upstream to catch the silt from Tenaya Creek. Environmentalists stopped this in 1971, allowing the lake's normal successional processes to occur. Now, silt from Tenaya Creek builds a delta at the east end of the lake, and in late summer and fall during low-water years, it becomes "Mirror Meadow."

As the lake shallows, sedges and grasses take root in the wet soil, followed quickly by willows. Over time, the combined transpiration of herbs and shrubs will lessen soil saturation, making it more amenable for aspens, whose seeds require a great deal of moisture to get started. As the water table drops, lodgepole pine seeds will germinate and root, along with colorful drifts of wildflowers.

When I arrive, the boulders ballasting the end of the lake are alive with children, a confetti of colors and voices, a Dufy landscape come to life. Water sparkles, trees leaf, children holler, babies chortle, birds sing, all's right with the world.

* * *

Another day, another hike, but at the trailhead here my heart sinks: there must be close to a hundred people. The hikes to Vernal and Nevada Falls are among the most popular in Yosemite, even in the off season, and up to two hundred thousand tourists a year here can create pedestrian jams on popular weekends. Blessedly, today the landscape swallows them up as people string out along the trail, each at his or her own pace.

I rest for a while at the foot of Vernal Fall, braced on a steep grassy slope ballasted with huge, sharp-edged boulders. Watching a waterfall is as fascinating as watching a campfire. Water pulses over the lip of the cliff and hurls downward as if someone threw it over in buckets, creating a series of white swags that blow across the rock face with pulsating hisses. Narrow diagonal ledges snag the pulses and splinter them into white lines as energized as jellyfish tentacles. When the wind takes the waterfall and turns it into gauze, it smokes across the shiny bronze and copper rock, lace curtains in an open window lifted by a breeze, shimmering with rainbows that wax and wane.

To ascend beyond Vernal Fall, one negotiates steep, rocky, leg-stretcher steps, iced with a skim of algae. Originally there was a slippery ladder up the old "Mist Trail," before steps were placed there in 1897. Helen Hunt Jackson, writing in *Bits of Travel at Home* in 1878,

Moss and lichen carpet a decaying stump of cedar. The tiny fruiting bodies of the moss are capsules atop threadlike stems, enclosing spores that will be loosed on a wilding wind when ripe.

Near Olmsted Point, a granite pool echoes the trees and boulders surrounding it.

Looking southwest from a wildflower meadow below Mount Dana, one of the highest peaks on the eastern edge of the park, 12,117- foot Mammoth Peak towers over the Kuna Crest.

made the earlier climb sound terrifying, jumping from slick stone to treacherous log, blinded with spray. The climb, despite the steps, still gives this hiker pause.

On top, Emerald Pool fills a hollow in the granite ledge above Vernal Fall. The gray granite has been scrubbed so pale over the millennia that it looks like an area of cement. Framing the pool are magnificent sugar pines ("sugar" comes from the crystalline blobs of sap that ooze out of a bark wound). Scotsman David Douglas (after whom the Douglas fir is named) searched for the source of the large pine nuts he had seen Indians carry. When he found the foot-long cones they came from, he discovered sugar pines. Today, large sugar pines persist mainly in protected park areas. Because they are good timber trees, they have been largely cut in forests outside the park. Their boles rise tall, straight, and clean up to a hundred feet, and their extravagant cones hang like Christmas tree decorations at the end of the limbs.

On a ledge between Vernal and Nevada Falls, Albert Snow opened a small inn in 1870 called La Casa Nevada, a pleasant looking place with a short happy life. It burned in 1891. At times of high water flow during Snow's tenancy, the Merced overflowed into a gulch north of Nevada Fall. Loathe to lose any of the water, Snow rerouted the channel in an operation he called "fixing the falls" to ensure that it all poured into the main river.

Glacial ice was exceedingly thick here, perhaps more than two thousand feet. On the platform above Nevada Fall sits a granite boulder dimpled with feldspar crystals, whose flat planes flash in the sunlight. This boulder is a glacial erratic, clearly different from the country rock upon which it sits. Even the 700-foot thickness of the last glacier was sufficient to drag this boulder along. When the glacier retreated, it left the boulder here to be swept up by the next glacier that never came.

* * *

When John Muir worked in Hutchings's sawmill at Yosemite, he set his cabin next to Yosemite Creek and rerouted a thread of it to flow through a corner of his cabin so that he could hear it "sing and warble in low, sweet tones, delightful at night while I lay in bed." The first time I saw Yosemite Falls, they were dry, as Horace Greeley, editorial writer for the *New York Tribune*, saw them (although disappointed by the lack of water, Greeley otherwise found the scenery "unique and majestic"). But this spring day, Yosemite Creek is anything but soporific. With huge spring runoff, the falls rush and pound and give plenty of aural clues to their power and verve, as Ansel Adams perfectly described them, "booming in early summer flood."

Yosemite Falls, the highest cascade in North America, sometimes jettisons hundreds of thousands of gallons per minute over its lip. The entire falls drop 2,425 feet, the prodigious descent divided into Upper and Lower Falls and separated by a Middle Fall, which cascades 675 feet. The entire falls are fifteen times higher than Niagara. The lowest, with the shortest drop of 320 feet, is twice the height of Niagara. The falls formed, as did the other

FOLLOWING PAGES:

A rainbow rises over Mount Dana, casting a glow over the meadows beneath.

*S*ome granodiorite contains dark lenses of diorite, a darker and finer grained type of granite that contains very little quartz. The parallel placement of these dark lenses suggests that they may have been plastic when caught in the molten granite so that they were stretched and aligned as the granite cooled. At the top are thin layers of granite in the process of exfoliation.

major cascades in Yosemite, in hanging valleys of smaller tributary streams.

At the base of Lower Fall, huge boulders shatter the water into foaming white disarray. Lower Yosemite Fall, its sound of tearing cloth so loud that it almost blots out the more familiar sounds of flowing water, assaults me with a churning wind. Splash and mist intermix in a great bounding, bouncing fall, sound and fury interwoven, ruffles and flourishes of spray, pure untrammeled energy at work.

A trail only 3.5 miles long leads to the top—an hour's walk on the flat, elongated by the steepness of the trail to a three- or four-hour climb. Joseph LeConte, before the current multiswitchback trail was built, wrote of having to proceed on hands and knees because the surface was so smooth and steep and of descending on the seat of his pants. John Conway began the trail in 1873. Characteristic of early trails, the zigs and zags are short and many, 135 switchbacks to be exact, and the pivots at the corners range from extraordinarily to unreasonably steep. Conway completed the trail in four years and operated it as a toll path until he sold it to the state in 1885. Price tag: fifteen hundred dollars.

After climbing through woods with few open vistas, the trail breaks out at a graveled point. A hundred feet away the top fall ends on a rock shelf and begins its galloping cascades to the top of Lower Fall. Its parabolic leap of water through 1,430 feet of empty space explodes water on the rock below. Spray boils up into clouds that mist the air. Underneath the continuous tearing and shrieking din, sonorous booms resonate like notes on a bass viol thrummed below the level of hearable sound, felt rather than heard.

Alongside the path, manzanita bushes bloom, flourishing on these well-drained and stony slopes. Their small leathery leaves resist water loss, and if their tops are destroyed by fire, they resprout prolifically from the roots. On this sunshiny spring morning, they are packed with clusters of small pink bells that are in turn frequented by painted lady butterflies and several sizes of bees, turning the bush into a spinning, buzzing mass of sound and flurry.

On the top of the falls, the speckled granite ledge contains small, dark gray ovals of diorite, inlaid like a design of little black fish all schooling in one direction. Diorite is a granitic rock with very little quartz in it, much darker and finer grained than the lighter El Capitan granite surrounding it. The shape of the intrusions suggests that they were malleable when they were caught in magma, and their stretched shape and parallel placement could have been caused by movement within the magma that squeezed and aligned them.

From here the valley is a very, very long way down. Roads appear to be an eighth-inch wide, and a toy Ahwahnee Hotel seems to be the size of a sugar cube. Straight across is Mount Starr King, named for the Unitarian minister who wrote about Yosemite shortly after it was discovered. Starr King is an unusually symmetrical dome, well-rounded by exfoliation, its height 9,092 feet. Beneath it is the silver line of Nevada Fall. Closer is Glacier Point, whose sheer cliffs were shaved by glaciers. Sentinel Dome, still snow-covered, rises behind it.

What I remember most about this day is not the vista but that while climbing toward a high sheer wall, I heard a canyon wren sing that exquisite, descending trill that enhances almost every river canyon in the West.

* * *

One of the greatest hikes in the world is to the summit of North Dome at 7,542 feet. It has everything: steeps and flats, views and vistas, cloisters and larking streams, browsing deer, elegant sheltered woods, dramatic exposures, and enough effort to make you feel you earned this nonpareil view 3,751 feet above Yosemite Valley.

I nestle in on the top, dive-bombed by a hummingbird, to eat lunch beneath four Jeffrey pines, three smaller ones and one with a trunk thirty inches in diameter. None rise higher than ten feet, bonsaied by the wind. Named after John Jeffrey, a Scots gardener who collected plants and seeds in California and Oregon in 1852–53, Jeffrey pines replace ponderosas at higher elevations. A ponderosa pine smells resinous and "piney," whereas Jeffrey pine has a milder vanilla odor, and the spines of Jeffrey pinecones turn inward while those of ponderosa turn out and prick your hand.

I look across the valley to the massive flat face of Half Dome, which seems but an arm's reach away. Indian legend relates that Tis-sa'-sack entered Yosemite Valley with her great conical gathering basket, and being thirsty, she drank all the water in Mirror Lake before her husband, who was equally thirsty, arrived. Upon seeing what she'd done, he beat her, and she retaliated by throwing her basket at him. For their wickedness they were turned to stone: the husband became Half Dome, the wife became North Dome, and beside her, Basket Dome.

A *pool in the Olmsted Point area mirrors sunset.*

The rounded summit of North Dome is still exfoliating, slabs of rock piled on top of each other like oversized sand bags. Three fence lizards cavort over them and up the tree trunks. A pair of golden-mantled ground squirrels conduct quick forays in search of crumbs. Iron oxides stain the granite with pale apricot. Marblelike dikes of aplite, less easily eroded than the mother rock, crisscross the summit like lines tethering it to the ground. In the cracks bright rosy-pink penstemons bloom among mustards and mosses and tiny grasses. I settle my back against a boulder of sun-warmed granite and absorb this halcyon noontime.

This July morning a light breeze streaks the otherwise calm surface of an alpine tarn, then dies, leaving a polished, stainless-steel skin. Yesterday my eldest daughter, Susan, and I hiked eight miles into the Sierra, and twenty-five hundred feet up, into the northeast corner of the Yosemite backcountry wilderness, which brought us to this charming tarn nestled in a granite bowl at around ten thousand feet.

The sky brightens but does not heat; I enjoy earth's staging time, getting everything adjusted, ordered, before turning on the sun. By six, full sunlight stains a ridge ruddy at the far edge of the lake, reflecting in the water like warp-dyed silk. Then this July day begins with blazing trumpets of light. I've never been averse to a little glory before breakfast, and watching sunrise bestir this pond does it for me.

I suspect the real glories of Yosemite belong to the backpackers, the trudgers and trekkers, those who finish a strenuous climb and wait for their psyches to catch up, suffer a thunderstorm on an alpine fell, and most of all, let the night spirits seep into their sleep. The real glories of Yosemite belong to those who are comfortable with being uncomfortable, who know it's all right to be afraid, to be cold, wet, tired and hungry, to be euphoric and, on occasion, ecstatic.

More than 706,000 acres, over 94 percent of the park, is managed as wilderness and can never

El Capitan looms above the Merced River on the north side of Yosemite Valley.

TRUMPETS OF LIGHT

BACKCOUNTRY JOURNAL

be developed. A permit system applies to hikers and groups on horseback who plan to remain overnight, thus guaranteeing that hikers are not falling over one another or over-using one area. The park instituted a permit system because rangers counted almost five thousand camp-fire rings in the backcountry in 1972. The wilderness areas cope well so far with the 30 percent of the visitors who go there, perhaps because they are of a different outlook than the 3 million who jam into the valley and spend their time commuting between the stores at the Visitor Center and Curry Village.

* * *

Kerrick Meadow lies a little above nine thousand feet, depending on a knob here or a depression there. Laid between granite walls, the valley looks quilted in all shades of green. In this wet soil, wildflowers abound, and little apricot-colored day moths flutter up against my legs as I walk. A grasshopper with dark brown wings alights at right angles to the sun, then ratchets off again.

Like better-known Tuolumne Meadows, Kerrick Meadow has seasonally satu-rated soils that maintain a water table too high for trees to grow. The meadow bears the name of James D. Kerrick, who trailed sheep here around 1880. Most of its yearly precipitation, between thirty and fifty inches, falls as snow, and the growing season seldom lasts more than nine weeks.

In the middle of Kerrick Meadow, Rancheria Creek (also probably a name used by sheepmen) flutters its way downhill on a gradual gradient. A tinsel-ribbon of water, it pauses occasionally to spread into small pools at the outside of a meander or to nibble at a bank, in no hurry to get anywhere. According to the sandy, gravelly flats alongside, its channel at high runoff sometimes widens to fifty feet. Frost heaves that make the ground expand disturb plant roots and leave bare, gravelly patches that resemble shaven spots in the lush sedge meadow that bounds the stream. These active, top layers of soil discourage plant growth and allow only the most sturdy pioneer plants, those that can withstand the thaws and freezes that unceremo-niously assault their roots. These plants, like tiny daisies and lupines and bright pink pussy-paws, have many of the same adaptations as those that occupy alpine heights. They are small, close to the ground, and often densely furred with hairs.

Following the creek, Susan and I sometimes walk in a horse pack trail incised six to twelve inches below the surface. One horse concentrates more pounds per square inch and causes as much damage as twenty-five or more people. Vegetation and soil at camps where horses are tethered is impacted ten times as much as other camps, and meadows are grazed into mud. But horse use continues because it provides easier and longer access to the backcountry.

* * *

An ominous, dark cloud to the north sits astride the valley upstream. From my perch I watch the virga, filmy veils of rain, shred down out of it. The storm crawls like a tank,

Cathedral Rocks on the south side of Yosemite Valley shimmer in the Merced River.

FOLLOWING PAGES:

The river gallops down its channel through the Grand Canyon of the Tuolumne River.

filling the breadth of the canyon, scraping the granite, dragging against the ground, marching toward us with an overweening arrogance. It formed from heated air rising from the Central Valley, cooling as it rose and capturing enough moisture to form thunderheads. A grumbling, muttering-in-its-beard thunder beats on the granite as if it were a tympanum, an unmistakable announcement of intent. One minute the rocks are dry, then suddenly the downpour rushes off their flanks. It rains the rest of the afternoon. It rains all night.

In the morning, watching a tent fly dry is in the same category as watching a pot boil, and the need to wait legitimizes some morning lethargy. I return to yesterday's storm-watching perch. Handfuls of moisture hang in the air, swathing everything in lingering damp-ness. Tiny yellow monkeyflowers that would fit into a shirt button, plants maybe less than half an inch high, interweave in familial mats. The lip of each flower carries a drop of water that magnifies the red speckles of its throat.

Spiny gooseberry bushes sprawl across the slope behind me, double thorns on the stems. Indians used to burn Yosemite Valley just to encourage such berry bushes to sprout, for gooseberry and other berry plants come in quicker after a burn. In more recent times, howev-er, gooseberry and currant, shrubs of the genus *Ribes*, have been indicted as being an alternate host to white pine blister rust. Blister rust infections come in waves, usually when cool and moist conditions encourage spores to form in the gooseberry phase. For decades foresters killed gooseberries and currants to stop the spread until, in the 1960s, the rust infection was judged not to be such a threat after all. Scattered plants that escaped execution remain at higher altitude.

While I lodgepole- and monkeyflower-watch, the sun levitates seven inches above the rock rim across the valley, and as it rises above the mists, it paints sharp clear shadows. With it comes a light breeze. I check my sleeping bag. Contrary to my pessimistic expectation of a soggy sleeping bag forever, it is dry.

* * *

Another afternoon, Susan and I hike to a higher tarn. The gravel apron around it scrunches underfoot. The water is so cold and lacking in minerals, so "pure," that no algae grow in it, no plankton, no fish.

A path crosses behind the tarn beneath its source of supply, a snowbank plastered on the scooped-out slope of granite wall a quarter mile away. Thick sedges hide threads of water that don't show up until you step into them ankle deep. At the edge of one rivulet, I spy a little half-inch tan frog with a black stripe through its eye. It hops off through skyscraper sedge—a Pacific tree frog.

After crossing the meadow, the path starts up to a divide. When it turns down again, I leave it to scramble up a bare talus slope. I intend to go only partway to a dark rocky ridge on the horizon, but it's a Pied Piper landscape, calling me through one more rock door-way, up one more rise to one more interesting plant, one more different kind of rock. I follow

Evening light softens the granite contours above the Glen Aulin area.

Leidig Meadow, bracketed between Sentinel Rock and Three Brothers in Yosemite Valley, floods in wet springs and floats springtime images.

FOLLOWING PAGES:

T*he Tuolumne River*

cascades down a steep

reach.

meekly, hypnotized, into the severe, sculptural spaces of one of my favorite places, the alpine zone.

The dark rock outcropping is so splintered into silver dollar pieces that it clinks as I walk through it. The metamorphic hornfels shatters into smaller chips than granite, contains more minerals, and is darker and more heat absorbent than the granites, an assist to plants at cold altitudes. This outcrop is a leftover from the metamorphics that once completely covered the mountains. Surrounded above, below, and alongside by pearly gray granite, the dark rock stands out as powerfully as the clenched fist in a Rodin sculpture.

Sun shines 354 days (almost 97 percent) of the year up here, but at the same time, it can freeze any night. Wind shaves the ground like a straight razor. Precipitation drops to less than half an inch in July, with August even drier. What little moisture there is comes from melting snow.

Plants up here tend to be low cushions and mats, white phlox and pink moss campion, tiny buckwheats and little lupines, all of which withstand freezing and can photosynthesize at lower temperatures than plants of lower altitude. Lupine is heavily furred, an adaptation that lessens water loss and insulates the plant against evaporation, solar radiation, and cold. Like most alpine plants, lupines are perennials. Most annuals do not have time enough to sprout, grow, flower, and set seed in the abbreviated growing season. Many species grow for a decade or more before they store up enough energy to flower. Some reproduce by vegetative means, which gives new plants a better start and an assured source of nourishment.

Plants here take root in nearly sterile, pulverized granite, chips of feldspar, quartz bits, and sparkly flecks of mica, with not enough organic matter in it to deserve the designation of "soil." With cold temperatures and few plants, humus neither forms nor stays put on steep alpine talus slopes. Red heather, full of pink blossoms, espaliers across a rock face, preferring the acids provided by disintegrating granites. White heather nestles against a pegmatite dike full of big handsome feldspar crystals. By growing on a slight incline facing the sun, the heather receives half again more heat and light than if it grew on the flat. Creeping mats of magenta penstemons and alpine sorrel always grow along the downhill edge of rocks, capturing the runoff moisture. A dainty sandwort raises little starlike blooms with five rosy stamens hovering above five starched white petals.

Downhill, puffy patches of flake lichen tint the ground an odd and distinctive bluish-gray. Flake lichen thrives within a growing season of seven to twelve weeks and signals that this area holds snow late into the summer. Snow is more protector than growth-stopper at high altitude. On days when the windchill factor may drop to minus forty degrees, snow shields alpine plants from a brutal buffeting.

Upslope, narrow rivulets seep out from under a raggedy patch of snow shining like tinsel. How does a snowfield die? Rather ignominiously, I'm afraid. A beautiful expanse of pristine white snow becomes crusted with dirt and dust, its surface porous and granulated from freezing over each night, melting each day, looking like Japanese rice paper with pine needles, willow leaves, and other plant bits encased in it. These absorb enough heat to sink in a quarter inch, blackening and imbedding themselves in a bezel of snow. Pink algae colors teacup-sized hollows. No longer big enough to sluice a stream of icy water downhill, the snowfield drips like a dozen leaky faucets. It languishes, passing away from sun disease, a fragile Camille, a wan Traviata, a doomed Mimi, dying with operatic slowness. It goes out neither with a bang nor a whimper—just a tiny liquid tinkling, its requiem the soprano of a mosquito.

As I start down, a muted clucking comes from nearby, stops, resumes. As quietly as possible, I slip out binoculars and wait. About ten feet away, close enough to see the red line above its eye, I spot a handsome male ptarmigan. About the size of a chicken, ptarmigan are an instance of Bergmann's Rule in action: creatures of cold climates tend to be larger than tropical animals, since size gives a better relationship of volume to exposed area and makes it easier to retain body heat. The bird proceeds with considerable dignity up a boulder face, snapping at flower heads, ambling up the rock to a small shady overhang. Speckled brown and white on back and wings, he blends into the dappled light under the overhang. In winter, ptarmigan turn totally white, with extravagant white pantaloons, a heavy feathering that gives extra insulation to legs and feet. Males winter above timberline, the only birds to do so in this stringent climate.

* * *

Later, after loping down a rain-greased talus slope with nothing taller in sight than I, lightning flashing and thunder banging simultaneously, a hastily donned poncho flapping, boots and pants soaked and hands stiff with cold, I finally reach a lower flat and hunker down. Just when I don't think I can get any wetter or any colder comes the hail. Stinging, petits-pois-sized pellets insinuate themselves into every crease of my poncho, fill the puddles around my feet, and leave windrows around the rocks. What isn't already wet gets wet—soaking, irrevocably, irretrievably wet, wet from the outside and through to the other side.

At the far edge of misery a pale sun appears, not a moment too soon for this huddled mass of dripping, dirty laundry, with runny nose and squishy boots. As the last BBs of hail clear the valley, a rainbow, an incredible swatch of color, materializes against the gloomy clouds. Not your same old arch but a rectangular banner broadcasting its blazing spectrum of color, it flutters out from under a mass of clouds in the southwest sky, undulating like northern lights.

That evening, tent pitched and trenched, clothes dry, and the comforts reestablished, I pull out the obligatory "ice-cream-and-cake-and-candle," a day-late birthday celebration for Susan: two small, slightly squashed cupcakes, one pink candle, and a packet of freeze-dried ice cream that tastes like ice cream even if it isn't cold. What's a mother for? We raise a toast of freshly filtered, very cold stream water to the pleasures of wilderness. My cup runneth over.

* * *

Waning sun, shuttered behind a pure white cloud, traces its rim with eye-blinding incandescence. Wisps of clouds radiate outward from Sawtooth Ridge like the gold rays in a baroque sculpture. Tall, dark green, narrow triangles of trees rise against a backdrop of white granite beneath a deep blue Sierran sky: unmistakably Yosemite. The sky chills to an aquamarine of limpid clarity and transparency, a cut-crystal atmosphere, before it deepens to navy blue in which stars begin to glint.

Shadows inch up the last sunlit face. With my sketchbook on my lap, I recall James D. Smillie, who published *Yellowstone to Yosemite: Early Adventures in the Mountain West* in 1872. Smillie wrote and illustrated his summer in Yosemite and with an artist's eye noted that Yosemite's granites, being so pale, are exceptionally responsive to changes in atmosphere. At sunset, he wrote that

> . . .*they glow with a ruddy light, that is slowly extinguished by the upcreeping shadows of night, until the highest point flames for one moment, then dies, ashy pale, under the glory that is lifted to the sky above. Then the cold moon tips with silver those giant, sleeping forms, and by its growing light I cleared my palette, and closed the box upon my last study of the Yosemite and Sierras.*

* * *

I unfold my map for the last time to check our route out tomorrow. Now its folds are worn, its edges shredded. Well-used indeed. Miles calculated, elevation lines counted, meadows walked, streams crossed, heights climbed. Now, when I trace with my finger where we've been on this trip, the map lines segue into images of clumps of pines or shining tarns or mellifluous meadows. That green spot here was full of flowers and butterflies, and that blue line there was a booming waterfall, sweetened with bird song. Those concentric ruffled circles describe the top of a dome on which I stood, those dotted blue lines the beginning of a stream that wiggled downslope.

No longer is the map two-dimensional. It is composed of height of ponderosa, breadth of valley, depth of stream, wintertime, summertime, springtime, autumntime, the vanilla smell of Jeffrey pine, the gritty feel of granite, the puckery taste of alpine sorrel, the unexpected song of a canyon wren, the senses of time, the waterfalls of the mind.

ACKNOWLEDGMENTS

Without two marvelous hiking companions, Susan Zwinger and Sarah Rabkin, I could not have seen all of Yosemite that I did. They gave generously of their time and spirit, and I hope they know how much I appreciate it.

My thanks also to Ginger Burley, Michael Ross, and George Wuerthner, who shared information about Yosemite, as well as to Wayne Merry, who was on the first climb of El Capitan, to King Huber for his critical eye, and to Page Stegner, for whom the park is backyard. Appreciation to Timilou Rixon, who helped to form the manuscript and provided insights, as she has done for many years and many books, and to Susan Wels, a meticulous and patient editor. Tom Lewis and Nancy Cash of Tehabi Books were pure joy to work with.

And a quiet and final thank you to Carl Sharsmith who, more than twenty years ago when I was working on *Land Above the Trees*, introduced me to the alpine tundra on the Dana Plateau. His joy and dedication to the alpine world live on.

ANN ZWINGER
Colorado Springs, Colorado

Special thanks to Keith Walklet for smoothing the way and, who along with Annette Bottaro-Walklet, Michael Frye, and Claudia Welsh (of the Ansel Adams Gallery), shared their experience and love of Yosemite; to Dorothy LeVain and Barbara Costello of the Curry Company for keeping track of my film shipments; to Beverly Lassister for trail tips; and to the stables crew of Eric Van Rader.

Thanks also to Serena Strelitz, my backcountry assistant who never complained about hiking the same trail three times for that special shot; to Jane and Ed Hawkins who kept me supplied with fresh fish and Yosemite stories; to my business partner Warren Cook and our studio librarian, Cyndy Geier; to Christy Jewell and the crew at A & I Color Labs; and finally to the bear in the backcountry whose beautiful face filled my light beam at 2:30 A.M. It was the magic moment of my time in Yosemite.

KATHLEEN NORRIS COOK
Ouray, Colorado

INDEX

Abrams, William Penn, 14, 17
Adams, Ansel, 17, 109
Ahwahnee Meadow, 98
Ahwahneechee Indians,
    14, 17,59, 122
Algae, 104, 129
Alpine plants, 128
Alpine tarns, 117, 122
Alpine zone, 73, 97, 128
American pipits, 97
Amphibians, 79, 87, 92
Ants, 80
Aplite, 114
Arêtes, 38
Ayres, Thomas, 25, 26

Basket Dome, 114
Batholiths, 36
Bees, 80, 84, 113
Beetles, 67, 80
Belding's ground squirrels, 79, 88
Bierdstadt, Albert, 26
Big Trees, 25, 60, 62-63, 67, 69, 75,
    80, 87
Bighorn sheep, 99
Black bears, 32, 92, 93, 96, 98, 99
Black oaks, 17, 59-60, 74, 87, 103
Bridalveil Fall, 17, 50, 53
Bridalveil Meadow, 48
Buckeye Pass, 97
Bunnell, Lt. Lafayette H.,
    17, 25, 26, 53, 74
Butterflies, 80, 84, 113

Canyon wrens, 114
Chickarees, 67, 72, 73, 79, 92
Chief Tenaya, 17
Chipmunks, 72, 73, 92
Clark's nutcrackers, 72, 73, 79-80,
    95, 97, 98
Climbing/climbers, 53
Concessionaires, 30, 33, 98-99
Conway, John, 14, 113
Corn lilies, 69
Coyotes, 93, 96, 99
Currants, 122
Curry, David C. and Jennie, 32
Daisies, 84, 118
Diorite intrusions,113

Dippers, 79, 95, 97
Domes, 14, 37-38, 44
Douglas, David, 109
Douglas firs, 25, 103, 109
Droughts, 60, 62

El Capitan, 48, 50, 53, 113
Emerald Pool, 109
Exfoliation, 37-38, 48, 114
Exotic species, 99

Fifth U.S. Cavalry, 30, 32, 60
Firefalls, 14
Fires/fire suppression, 59, 60, 67, 69,
    75
Flies, 80, 84, 87, 92
Four-Mile Trail, 14
Fungus, 63, 67

Galls, 87
General Sherman Tree, 63
Ghost forests, 87
Giardia lamblia, 80
Glacial evidence, 48
Glacial erratics, 48, 109
Glacier Point, 13, 14, 32, 48, 113
Glaciers, glaciation, 37, 38, 40, 48,
    50, 109
Gooseberries, 122
Grand Canyon of the Tuolumne, 44
Granite(s), 35, 36, 37, 48, 50, 53
    114, 128
Gray-crowned rosy finches, 97
Greeley, Horace, 13, 26, 109
Grizzly bears, 14, 99
Grizzly Giant, 69
Ground squirrels, 92, 114

Half Dome,13, 14, 38, 48, 53, 104,
    114
Hanging valleys, 50, 113
Harding, Warren, 53
Hawks, 97
Hetchy Hetchy Reservoir, 44
Hornfels, 36, 128
Horse trails, 118
Hotels, 14, 25
Hummingbirds, 114
Hutchings, James, 25, 30, 32, 109

Incense cedars, 25, 59, 60, 69, 74
Insects, 79, 80, 92, 98
Inspiration Points, 17, 25, 87

Jackson, Helen Hunt, 104
Jeffrey, John, 114
Jeffrey pines, 25, 59, 114
Johnson, Robert Underwood, 72
Joints, jointing, 37, 38
Juncos, 79, 97

Kerrick Meadow, 118
Kerrick, James D., 118

LeConte, Joseph, 26, 38, 72, 113
Lembert Dome, 38, 48
Leonard, Zenas, 62
Lichens, 25, 128
Lincoln, Abraham, 25, 30
Little pink elephants, 84
Lizards, 79, 92, 114
Lodgepole pines, 25, 72, 73, 75, 87,
    104
Lupines, 118, 128

Manzanitas, 87, 113
Mariposa Grove, 25, 30, 32, 62, 67,
    69, 72
Mariposa Battalion, 14, 17
Marmots, 79
Matterhorn Peak, 38
McCauley, James, 14
Merced River, 14, 44, 45, 50, 62,
    109
Merced Peak, 38
Metamorphic rocks, 36, 37, 128
Mice, 73, 92
Mirror Lake/Road, 48, 103, 104, 114
Mirror Meadow, 104
Mist Trail, 104
Mountain House, 14
Mixed conifer forest, 25, 60, 72,   79
Monkeyflowers, 122
Moraines, 48, 50, 103
Moran, Thomas, 17
Mosses, 60, 103-104, 114
Moths, 84, 87, 118
Mount Dana, 37
Mount Maclure, 44, 50

Mount Starr King, 113
Mount Lyell, 44, 50, 53, 59, 88
Mount Lyell salamanders, 88
Mountain lions, 32, 79, 92, 99
Mountain hemlocks, 72
Muir, John, 17, 26, 30, 32, 38, 60,
    72, 73, 87, 97, 109
Mule deer, 92, 99

National Park Act, 30
Nevada Fall, 50, 104, 109, 113
North Dome, 38, 79, 97, 104, 114
Nunataks, 73

Olmsted, Frederick Law, 30, 33
Owens Valley earthquake, 50

Pacific tree frogs, 87, 95, 122
Penstemons, 84, 114, 128
Piute Creek, 50
Plutons, 36, 53
Pohono Trail, 14, 17
Polly Dome, 48
Ponderosa pines, 59, 60, 67, 74, 114
Porcupines, 99
Precipitation/rainfall, 60, 62, 63,
    118, 122, 128
Prescribed burns, 67, 69
Ptarmigan, 97, 129
Pussypaws, 118

Rancheria Creek, 97, 118
Red-winged blackbird, 93
Riparian vegetation zone, 103
Roches moutonnées, 38
Rock Island Pass, 84
Rockfalls/rockslides, 50
Roosevelt, Theodore,26, 32

Savage, James, 14, 17
Sawtooth Ridge, 130
Sedges, 72, 73, 122
Sentinel Dome, 35, 37, 38, 114
Sequoiadendron giganteum, 62, 69
Sheep herders, 32, 72
Sheep, domestic/overgrazing, 32, 72,
    99
Shrews, 79, 92, 93
Sierra Club, 26, 30

Sierra Nevada, 13, 25, 30, 36, 37,
    38, 44, 60, 63, 88
Slide Mountain, 50
Smillie, James D., 130
Snow plants, 69
Snow, Albert, 109
Spalling, 48, 50
Spotted owls, 95
Starr King, 113
Subalpine meadows, 69, 75, 87
Sugar pines, 25, 59, 75, 109

Tectonic plates, 35, 36
Tenaya Valley, 104
Tenaya Creek, 103, 104
Tioga Pass, 69
Tis-sa'-sack, 114
Tuolumne River, 44, 45, 62, 69
Tuolumne Meadows, 69, 72, 118
Tuolumne Cascades, 69

U. S. Army, 17
U-shaped valleys, 48

V-shaped valleys, 37, 48
Vernal Fall, 50, 104, 109
Volcanoes, 36, 37

Walker, Joseph, 62
Warblers, 97, 98
Water ouzels. See dippers
Waterfalls, 13, 14, 17, 50
Wawona, 25, 30, 60
Weasels, 93, 97
White pine blister rust, 122
White firs, 59, 60, 69, 74
White-crowned sparrows, 97-98
Whitebark pines, 59, 72, 73, 75, 95
Whitmore, George, 53
Whitney, Josiah, 38
Wilderness areas/permits, 45,
    117-118
Willows, 59, 97, 104
Wolverines, 99

Yosemite Creek, 32, 109
Yosemite toads, 87
Yosemite Commission, 30
Yosemite Falls, 50, 53, 79, 88, 109,
    113

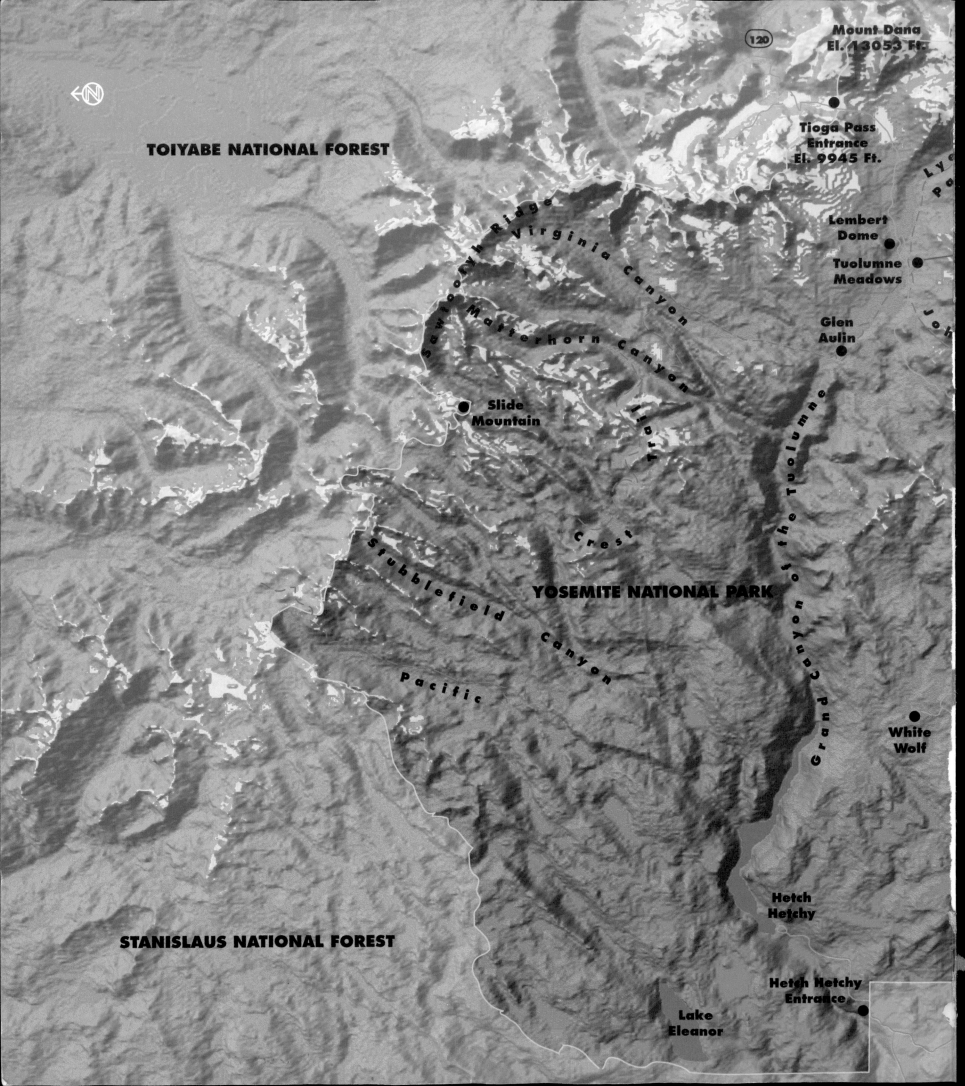